Adaptation, Learning, and Optimization

Volume 26

The role of adaptation, learning and optimization are becoming increasingly essential and intertwined. The capability of a system to adapt either through modification of its physiological structure or via some revalidation process of internal mechanisms that directly dictate the response or behavior is crucial in many real world applications. Optimization lies at the heart of most machine learning approaches while learning and optimization are two primary means to effect adaptation in various forms. They usually involve computational processes incorporated within the system that trigger parametric updating and knowledge or model enhancement, giving rise to progressive improvement. This book series serves as a channel to consolidate work related to topics linked to adaptation, learning and optimization in systems and structures. Topics covered under this series include:

- complex adaptive systems including evolutionary computation, memetic computing, swarm intelligence, neural networks, fuzzy systems, tabu search, simulated annealing, etc.
- machine learning, data mining & mathematical programming
- hybridization of techniques that span across artificial intelligence and computational intelligence for synergistic alliance of strategies for problem-solving.
- aspects of adaptation in robotics
- agent-based computing
- autonomic/pervasive computing
- dynamic optimization/learning in noisy and uncertain environment
- systemic alliance of stochastic and conventional search techniques
- all aspects of adaptations in man-machine systems.

This book series bridges the dichotomy of modern and conventional mathematical and heuristic/meta-heuristics approaches to bring about effective adaptation, learning and optimization. It propels the maxim that the old and the new can come together and be combined synergistically to scale new heights in problem-solving. To reach such a level, numerous research issues will emerge and researchers will find the book series a convenient medium to track the progresses made.

Indexed by SCOPUS, zbMATH, SCImago.

More information about this series at https://link.springer.com/bookseries/8335

Kyle Robert Harrison · Saber Elsayed ·
Ivan Leonidovich Garanovich · Terence Weir ·
Sharon G. Boswell · Ruhul Amin Sarker
Editors

Evolutionary and Memetic Computing for Project Portfolio Selection and Scheduling

Springer

Editors
Kyle Robert Harrison
UNSW Canberra
Canberra, ACT, Australia

Saber Elsayed
UNSW Canberra
Canberra, ACT, Australia

Ivan Leonidovich Garanovich
Department of Defence
Defence Science and Technology Group
Canberra, ACT, Australia

Terence Weir
Department of Defence
Defence Science and Technology Group
Canberra, ACT, Australia

Sharon G. Boswell
Department of Defence
Defence Science and Technology Group
Canberra, ACT, Australia

Ruhul Amin Sarker
UNSW Canberra
Canberra, ACT, Australia

ISSN 1867-4534 ISSN 1867-4542 (electronic)
Adaptation, Learning, and Optimization
ISBN 978-3-030-88314-0 ISBN 978-3-030-88315-7 (eBook)
https://doi.org/10.1007/978-3-030-88315-7

This Springer imprint is published by the registered company Springer Nature Switzerland AG
The registered company address is: Gewerbestrasse 11, 6330 Cham, Switzerland

Preface

Many organizations face the challenge of selecting and scheduling a subset of available projects subject to various resource and operational constraints. Formally, this problem is known as the project portfolio selection and scheduling problem (PPSSP). It is well known that the PPSSP is an NP-hard problem, thus, there is no known polynomial algorithm for it. Despite the complexity associated with solving the PPSSP, many traditional approaches to this problem make use of exact solvers. While exact solvers provide definitive optimal solutions, they quickly become prohibitively expensive in terms of computation time when the problem size is increased. In contrast, evolutionary and memetic computing afford the capability for autonomous heuristic approaches and expert knowledge to be combined, thereby they provide an efficient means for high-quality approximation solutions to be attained. As such, these approaches can provide near real-time decision support information for portfolio design that can be used to augment and improve existing human-centric strategic decision-making processes.

This edited book provides the reader with a broad overview of the PPSSP, its associated challenges, and approaches to addressing the problem using evolutionary and memetic computing. Contributions were welcomed from distinguished researchers and industry practitioners that highlight the state of the art and recent trends in memetic computing and evolutionary computation for the PPSSP in any domain. Each chapter published in this book has been peer-reviewed by a minimum of two experts in the field. We gratefully acknowledge the reviewers, listed below, for their valuable feedback and constructive comments on the contributed chapters, which have greatly improved this book.

- Ripon Chakrabortty, UNSW Canberra, Australia
- Mohammad Humyun Fuad Rahman, UNSW Canberra, Australia
- Amany AKL, UNSW Canberra, Australia
- Ali Ahrari, UNSW Canberra, Australia
- Noha Hamza, UNSW Canberra, Australia
- Mohamed Radwan, UNSW Canberra, Australia
- Daryl Essam, UNSW Canberra, Australia

- Marjia Haque, UNSW Canberra, Australia
- Sanjoy Paul, University of Technology Sydney, Australia
- Danilo A. Milhomem, Pontifícia Universidade Católica de Goiás (Catholic University of Goiás), Brazil

We would like to express our gratitude to the staff and editorial team of the Adaptation, Learning, and Optimization book series as well as the publisher, Springer Nature Switzerland AG. Specifically, we would like to acknowledge the professional support provided by the Project Coordinators Ms. Jennifer Sweety Johnson and Mr. Suresh Dharmalingam, as well as the Editorial Director Dr. Thomas Ditzinger at Springer Nature. A special note of thanks must also go to the Defence Science and Technology Group (DSTG), Australian Department of Defence, for their support of this edited volume.

Canberra, Australia
July 2021

Kyle Robert Harrison
Saber Elsayed
Ivan Leonidovich Garanovich
Terence Weir
Sharon G. Boswell
Ruhul Amin Sarker

Contents

Evolutionary and Memetic Computing for Project Portfolio Selection and Scheduling: An Introduction 1
Kyle Robert Harrison, Ivan L. Garanovich, Terence Weir, Sharon G. Boswell, Saber M. Elsayed, and Ruhul Amin Sarker

Evolutionary Approaches for Project Portfolio Optimization: An Overview 9
Ruhul Amin Sarker, Kyle Robert Harrison, and Saber M. Elsayed

An Introduction to Evolutionary and Memetic Algorithms for Parameter Optimization 37
Ali Ahrari and Daryl Essam

An Overall Characterization of the Project Portfolio Optimization Problem and an Approach Based on Evolutionary Algorithms to Address It 65
Eduardo Fernández, Efrain Solares, Carlos A. Coello Coello, and Victor De-León-Gómez

A New Model for the Project Portfolio Selection and Scheduling Problem with Defence Capability Options 89
Kyle Robert Harrison, Saber M. Elsayed, Ivan L. Garanovich, Terence Weir, Sharon G. Boswell, and Ruhul Amin Sarker

Analysis of New Approaches Used in Portfolio Optimization: A Systematic Literature Review 125
Danilo A. Milhomem and Maria José Pereira Dantas

A Temporal Knapsack Approach to Defence Portfolio Selection 159
Gregory Calbert, Terence Weir, Ivan L. Garanovich, and Carlos C. N. Kuhn

A Decision Support System for Planning Portfolios of Supply Chain Improvement Projects in the Semiconductor Industry 193
Rainer Kolisch and Thomas Fliedner

Index 213

Evolutionary and Memetic Computing for Project Portfolio Selection and Scheduling: An Introduction

Kyle Robert Harrison, Ivan L. Garanovich, Terence Weir, Sharon G. Boswell, Saber M. Elsayed, and Ruhul Amin Sarker

Abstract The problem of selecting and scheduling a subset of projects is a complex task faced by many organizations. This problem is referred to as the project portfolio selection and scheduling problem (PPSSP). In nearly all cases, a budgetary constraint limits the number of projects that can be selected. Furthermore, complex operational constraints are often present as well. Hence, the selection of the optimal set of projects is a challenging task, which is further complicated by the necessity also to determine the scheduling of their implementation. In many cases, intelligent techniques inspired by evolutionary processes are employed to address such problems. This introductory chapter provides a short overview of the PPSSP and the relevant solution methodologies, providing context for the remainder of this book. A brief description of each contributed chapter is also provided.

K. R. Harrison (✉) · S. M. Elsayed · R. A. Sarker
University of New South Wales Canberra at the Australian Defence Force Academy, Canberra, Australia
e-mail: kyle.harrison@unsw.edu.au

S. M. Elsayed
e-mail: s.elsayed@unsw.edu.au

R. A. Sarker
e-mail: r.sarker@adfa.edu.au

I. L. Garanovich · T. Weir · S. G. Boswell
Joint Operations and Analysis Division, Defence Science and Technology Group, Department of Defence, Canberra, Australia
e-mail: ivan.garanovich@dst.defence.gov.au

T. Weir
e-mail: terence.weir@dst.defence.gov.au

S. G. Boswell
e-mail: sharon.boswell@dst.defence.gov.au

K. R. Harrison et al., *Evolutionary and Memetic Computing for Project Portfolio Selection and Scheduling*, Adaptation, Learning, and Optimization 26,
https://doi.org/10.1007/978-3-030-88315-7_1

1 Introduction

Many people tend to assume that the term "portfolio" refers to a financial portfolio. This is likely a result of the seminal 1952 study of Markowitz [9], along with the fact that this domain is the most likely to be encountered by a layperson. In simple terms, a portfolio just represents a collection of things and can be encountered in any number of different application areas. For example, (i) in financial markets, a portfolio can represent a collection of financial investments, such as stocks, commodities, cash, and cash equivalents, (ii) in a corporate environment, it can represent a collection of products, services, and/or achievements of the company, and (iii) in any organization dealing with multiple projects, it can represent a collection of projects that share one or more resources.

Portfolio optimization is the process of selecting the best portfolio out of the set of all potential portfolios being considered, according to some objective(s). In a financial context, used for demonstration purposes, the primary goal is to select a subset of investment options (i.e., allocate the available assets to options) that provides the maximal expected return and does not violate the asset limitation. In some cases, additional objectives like minimizing the risk of return and minimizing the variance of returns are also considered. That means the optimization problem can be either single- or multi-objective. Once the investment decision is made, the investments may continue for years or can be revised depending on changes in the market conditions, assets, and/or the investment strategy.

A financial portfolio optimization strategy usually applies the concept of diversification, which means investing in a wide variety of asset types and classes. Diversification across asset classes is basically a risk-mitigation strategy. Here, the risk is defined as the chance that the investor will lose money on an investment or would not see the returns that are expected. However, investors also measure risk through volatility, which refers to the likelihood that an asset's price will change significantly. To deal with risk in the optimization process, additional constraints are usually imposed on the investment amount for each risk group (for example, high, medium, and low risk options). As discussed earlier, risk minimization can be considered as an objective. Analogous scenarios and problem descriptions arise in other domains.

In this introductory chapter, a brief discussion of the various general aspects of the PPSSP is presented, along with an overview of different solution methodologies. A brief summary of the contributed chapters is also presented.

2 Problem Formulation

In project portfolio optimization, the basic problem is to select a subset of projects, from a defined list, in order to maximize some measure of return or any other suitable objective measure, while satisfying the budget constraint. Project portfolio optimization is usually used as a planning tool by organizations that need to allocate funding to different projects and execute them as part of their organizational function. For example, this problem arises in such areas as:

- Public and Social projects [1],
- Software/IT [8],
- R&D and Production [10],
- Construction and Infrastructure [12],
- Investment [13], and
- Defense [4, 5, 7, 16].

Depending on the type of organization, the projects can be of different types and have differing characteristics. For example, the projects can be a mix of:

- small to large projects based on the budget, resources, and other technical requirements,
- short to long projects based on their duration of completion, and
- early to late start projects depending on their requirement within the organization.

When the start time of projects must also be decided during the optimization process, the problem becomes a combined resource allocation and scheduling problem known as the project portfolio selection and scheduling problem (PPSSP). In such problems, there may be a subset of the projects that must be selected, i.e., they are mandatory. The most straightforward approach to deal with mandatory projects is to exclude them from the optimization process and adjust the resources accordingly. Other common conditional relations that may arise between various projects are:

- precedence relationships: projects that cannot be executed until other specific projects have been completed,
- alternative projects: projects that serve the same purpose, such that only one should be executed,
- mutually exclusive projects: projects that cannot be executed simultaneously,
- mutually inclusive projects: projects that must be executed simultaneously or jointly,
- priority: some projects may have a higher priority compared to others, and
- project grouping: the projects may be divided into groups, and different budget limits and/or conditions may be applied to different groups.

The problem may then be formulated to have one or more objectives, depending on the goal of the organization. In the literature, the problem is defined in many different ways (considering a subset of the constraints and conditions discussed above). Additionally, uncertainty in data, missing data, and changing data with time are a few of the practical issues that should be considered. As a result, there are a variety of approaches, such as mathematical modeling with exact algorithms, and heuristics and evolutionary algorithms. In the next section, a brief discussion of solution methodologies is given.

3 Solution Methodologies

Just as there are many different formulations of portfolio optimization problems, there are various different optimization techniques that can be employed to address them. This section gives a brief introduction to three broad classes of optimization approaches that are commonly applied to PPSSPs.

3.1 *Mathematical Optimization*

In simple terms, the primary objective of an optimization problem is to find the set of input values that realize the maximum or minimum of some objective function, while adhering to a set of constraints. Clearly, there is an inherent mathematical underpinning to optimization. It should come as no surprise that many optimization approaches rely directly on mathematical modeling to aid in the decision making process. The use of mathematical derivation techniques to find the (feasible) optimum of an objective function is referred to as *conventional optimization*. These approaches leverage mathematical properties, such as gradient information, to derive an optimal solution for an optimization problem [14].

When addressing a problem using conventional optimization, it is important to consider the characteristics of the decision variables, objective function, and constraints. For example, the decision variables might be continuous or discrete, the objective function can be linear or non-linear, and the constraints may be convex or non-convex. It is well known that different problem characteristics require different optimization approaches with drastically different complexities. As an example, solving a linear program with real variables can be done in polynomial time, whereas solving an integer linear program is NP-hard—this implies that there is no known polynomial-time algorithm to provide solutions to integer linear problems, in general [6].

One of the most common approaches to address integer programs, using conventional optimization, is known as *branch and bound*. Branch and bound is a technique to recursively enumerate a search space by partitioning candidate solutions into subsets, referred to as *branches*. The best known upper and lower *bounds* for the objective value are maintained and used to discard branches that are incapable of improving on the (current) bound. This pruning strategy leads to a reduction in the number of solutions that must be enumerated, which can significantly lower the time needed to arrive at an optimal solution. However, given that solving integer linear programs is known to be NP-hard, branch and bound is still computationally infeasible for many real-world problems. One approach to mitigate the high computational burden is to introduce a relaxation (i.e., a simplification) of the model. The relaxed model can be used to identify feasible solutions much more rapidly, but these solutions are often sub-optimal with respect to the original (non-relaxed) model [15].

3.2 Evolutionary Computation

Evolutionary computation (EC) refers to an optimization paradigm that draws inspiration from the principles of Darwinian evolution. More specifically, EC techniques are stochastic, meta-heuristic search procedures that are characterized by three prominent features. Firstly, EC approaches are population-based search processes whereby the search simultaneously "evolves" a collection of candidate solutions. This is in contrast to conventional optimization approaches, which typically operate by performing an iterative improvement of a single solution. Secondly, EC approaches provide a mechanism to share and/or recombine information between individuals in the population. The exchange of genetic material is performed through genetic operators, such as mutation and crossover, which are used to produce new individuals for (possible) insertion into the population. Finally, EC approaches adopt a survival-of-the-fittest paradigm. A selection bias is employed in favor of better-fit individuals, thus providing an inherent pressure towards improving the overall suitability of the population as a whole.

Approaches from the field of EC are widely used to address optimization problems that are too challenging for conventional optimization. Notably, evolutionary approaches do not require a complete enumeration of the search space, thus they can be efficiently applied to problems that are computationally infeasible for conventional optimization. However, EC approaches do not provide any guarantee on the quality of the solutions they generate. Rather, they are used to find high-quality approximate solutions in significantly less time than conventional approaches. As a further benefit of EC approaches, in most cases only the fitness function is problem-specific, thereby providing a flexible approach that can be easily adapted to new problems [3]. Nonetheless, each of the components of an evolutionary approach can be carefully selected to craft a tailor-made optimizer for a particular problem [2].

3.3 Memetic Computing

Memetic computing refers to a computational paradigm that facilitates the sharing of information between different computational processes, i.e., the hybridization of two or more optimization approaches. Often, this manifests as the combination of a population-based meta-heuristic, such as an EC approach, with a local search process [11]. This hybrid algorithm is referred to as a memetic algorithm (MA). In many cases, the local search component is heavily inspired by the particular problem at hand and often draws inspiration from heuristic techniques employed in conventional optimization. Therefore, one can view memetic algorithms (MAs) as a hybridization of EC and conventional optimization approaches, such that they leverage the diverse, rapid search process of the former and the rigorous mathematical derivations of the latter.

4 Summary of Chapters

This edited volume consists of seven chapters and presents recent trends in project portfolio selection and scheduling, focusing on the evolutionary and memetic computing approaches used to address such problems. An outline of the chapters is provided in Table 1, while a brief summary of each chapter follows.

Chapter "Evolutionary Approaches for Project Portfolio Optimization: An Overview" presents a survey on the use of evolutionary approaches to address project portfolio optimization problems. The review focuses on three fundamental aspects, namely the application areas, the novelties found in their mathematical models, and the evolutionary approaches employed to solve the models.

Chapter "An Introduction to Evolutionary and Memetic Algorithms for Parameter Optimization" presents an introduction to the ideas and formulations of evolutionary algorithms (EAs). The chapter features a discussion of the advantages and disadvantages of EAs in comparison to conventional (mathematical) optimization methods. Finally, the chapter provides a brief introduction to memetic algorithms as an extension of EAs.

Chapter "An Overall Characterization of the Project Portfolio Optimization Problem and an Approach Based on Evolutionary Algorithms to Address It" provides a characterization of the project portfolio selection problem and proposes a formal problem statement that considers such features. It is argued that previously published approaches to project portfolio selection problem do not adequately address the entire complexity. In addition, the chapter provides insight into plausible components that evolutionary approaches should adopt to address the generalized problem.

Table 1 Summary of contributed chapters

Chapter	Chapter title	Domain
2	Evolutionary approaches for project portfolio optimization: an overview	General problems
3	An introduction to evolutionary and memetic algorithms for parameter optimization	General methodologies
4	An overall characterization of the project portfolio optimization problem and an approach based on evolutionary algorithms to address it	Social projects
5	A new model for project portfolio selection and scheduling with defence capability options	Defense projects
6	Analysis of new approaches used in portfolio optimization: a systematic literature review	Investment projects
7	A temporal knapsack approach to defence portfolio selection	Defense projects
8	A decision support system for planning portfolios of supply chain improvement projects in the semiconductor industry	Manufacturing projects

Chapter "A New Model for the Project Portfolio Selection and Scheduling Problem with Defence Capability Options" examines a real-world use case of the PPSSP in the defense sector. The chapter develops a new model to address the PPSSP in the context of the Australian Defence Force (ADF) capability development process. A custom-built heuristic approach is developed, with evolutionary approaches used to further refine the heuristic solutions.

Chapter "Analysis of New Approaches Used in Portfolio Optimization: A Systematic Literature Review" presents a systematic literature review of recent approaches used in portfolio optimization. This review examines and analyzes the solution methodologies, software implementation framework, model constraints, and technical analysis from a variety of studies, largely focusing on investment applications. This chapter also discusses the gaps identified in the literature and provides suggestions for future research.

Chapter "A Temporal Knapsack Approach to Defence Portfolio Selection" examines a second application of portfolio optimization in the defense sector. In this chapter, a set of candidate projects was first determined through wargaming and cost modeling. The "value" associated with each project was dictated by the year in which it commences, thereby incorporating a temporal aspect. This problem is formulated as a mixed-integer linear program (MILP), which resembles a temporal knapsack problem, and a methodology to solve it using Microsoft Excel™ is discussed.

The final chapter, Chap. "A Decision Support System for Planning Portfolios of Supply Chain Improvement Projects in the Semiconductor Industry", provides an example of the portfolio selection process in the manufacturing domain. This chapter describes the development of an integer programming model based on a multi-attribute portfolio value model. A discussion on how the model developed was used to build a decision system that would aid decision makers during committee meetings is given. Observations and insights from a real-world case study are presented.

5 Guide for Readers

This section provides a guide for readers to assist with navigating this book depending on the background and/or intentions of the reader.

For those not familiar with the field of EC and EAs, in general, we recommend to begin by reading Chap. "An Introduction to Evolutionary and Memetic Algorithms for Parameter Optimization". For those wanting an overview of the state-of-the-art in PPSSP literature, we recommend reading Chaps. "Evolutionary Approaches for Project Portfolio Optimization: An Overview", "An Overall Characterization of the Project Portfolio Optimization Problem and an Approach Based on Evolutionary Algorithms to Address It", and "Analysis of New Approaches Used in Portfolio Optimization: A Systematic Literature Review". For readers interested in chapters detailing the use of EC for portfolio selection and/or scheduling, we recommend Chaps. "Evolutionary Approaches for Project Portfolio Optimization: An Overview",

"An Overall Characterization of the Project Portfolio Optimization Problem and an Approach Based on Evolutionary Algorithms to Address It", and "A New Model for the Project Portfolio Selection and Scheduling Problem with Defence Capability Options". For readers interested in the use of conventional optimization techniques, we recommend reading Chaps. "A Temporal Knapsack Approach to Defence Portfolio Selection" and "A Decision Support System for Planning Portfolios of Supply Chain Improvement Projects in the Semiconductor Industry".

References

1. Arratia, M.N., Lopez I.F., Schaeffer, S., Cruz-Reyes, L.: Static R&D project portfolio selection in public organizations. Decis. Support Syst. **84**, 53–63 (2016)
2. Back, T., Hammel, U., Schwefel, H.P.: Evolutionary computation: comments on the history and current state. IEEE Trans. Evol. Comput. **1**(1), 3–17 (1997)
3. Eiben, A.E., Smith, J.: From evolutionary computation to the evolution of things. Nature **521**(7553), 476–482 (2015)
4. Garanovich, I.L., de Visser, G., Nguyen, M.T., Gill, A., Ween, A., Heseltine, T., Jiang, L., Taylor, R., Watson, J., Zadeh, H.S.: Capability portfolio management. In: Proceedings of the 2013 Systems Engineering and Test and Evaluation Conference, pp. 1–13. Canberra, Australia (2013)
5. Garanovich, I.L., de Visser, G., Nguyen, M.T., Gill, A., Ween, A., Heseltine, T., Jiang, L., Watson, J., Taylor, R., Tailby, D., Zadeh, H.S.: Management of interdependencies in defence capability portfolio. In: Piantadosi, J., Anderssen, R., Boland, J. (eds.) 20th International Congress on Modelling and Simulation, December, pp. 1012–1018. Modelling and Simulation Society of Australia and New Zealand (2013). https://doi.org/10.36334/modsim.2013.D1.garanovich
6. Genova, K., Guliashki, V.: Linear integer programming methods and approaches - a survey. Cybern. Inf. Technol. **11**(1), 3–25 (2011)
7. Harrison, K.R., Elsayed, S., Garanovich, I., Weir, T., Galister, M., Boswell, S., Taylor, R., Sarker, R.: Portfolio optimization for defence applications. IEEE Access **8**(1), 60152–60178 (2020)
8. Karimpour, R., Ruhe, G.: Evolutionary robust optimization for software product line scoping: an explorative study. Comput. Lang., Syst. Struct. **47**, 189–210 (2017)
9. Markowitz, H.: Portfolio selection. J. Financ. **7**(1), 77–91 (1952)
10. Medaglia, A.L., Graves, S.B., Ringuest, J.L.: A multiobjective evolutionary approach for linearly constrained project selection under uncertainty. Eur. J. Oper. Res. **179**(3), 869–894 (2007)
11. Neri, F., Cotta, C.: Memetic algorithms and memetic computing optimization: a literature review. Swarm Evol. Comput. **2**, 1–14 (2012)
12. Panayiotou, A., Medda, F.: Portfolio of infrastructure investments: analysis of European infrastructure. J. Infrastruct. Syst. **22**(3), 04016011 (2016)
13. Rather, A.M., Sastry, V.N., Agarwal, A.: Stock market prediction and Portfolio selection models: a survey. OPSEARCH **54**(3), 558–579 (2017)
14. Sarker, R.A., Newton, C.S.: Optimization Modelling: A Practical Approach, 1st edn. Taylor & Francis Group, Boca Raton (2008)
15. Schwefel, H.P.: Advantages (and disadvantages) of evolutionary computation over other approaches. Evolutionary Computation, chap. 1, pp. 20–22. CRC Press, Boca Raton (2000)
16. Zwikael, O., Garanovich, I.L., Weir, T., Gowlett, P., Aljoofri, S., Ween, A.: Identifying triggers for new projects and programs in the australian department of defence. In: Proceedings of the Australian Institute of Project Management National Conference, pp. 29–38. Australian Institute of Project Management (2018)

Evolutionary Approaches for Project Portfolio Optimization: An Overview

Ruhul Amin Sarker, Kyle Robert Harrison, and Saber M. Elsayed

Abstract This chapter presents a brief review of evolutionary approaches to address project portfolio optimization problems. This review focuses on three key aspects. Firstly, the problem descriptions are reviewed by grouping them into six broad application areas. Secondly, the key novelties found in their respective mathematical models are discussed. Finally, the evolutionary approaches for solving these models are reviewed. Key findings are also presented in tabular form for ease of reference and comparison.

1 Introduction

In project management, a project is defined as a set of activities (or tasks) that usually require different amounts of time and resources, must be accomplished in a specific order, and must achieve a combined goal [2]. A project portfolio is then a set of projects that share one or more common resources. A project portfolio decision problem is an upper level project selection process that is to select a set of projects from a long list of competing projects based on one or more objectives, goals, or criteria while satisfying one or more constraints [2, 19, 39, 40]. This problem is known as a project portfolio optimization problem or a project portfolio selection problem. There are a few steps in solving an optimization problem that includes problem identification, problem definition, mathematical representation of the problem (also known as formulating a mathematical model), solving the model (solution approach), result interpretation, and implementation [36]. This chapter aims to present a brief

R. A. Sarker (✉) · K. R. Harrison · S. M. Elsayed
University of New South Wales Canberra at the Australian Defence Force Academy, Canberra, Australia
e-mail: r.sarker@adfa.edu.au

K. R. Harrison
e-mail: kyle.harrison@unsw.edu.au

S. M. Elsayed
e-mail: s.elsayed@unsw.edu.au

K. R. Harrison et al., *Evolutionary and Memetic Computing for Project Portfolio Selection and Scheduling*, Adaptation, Learning, and Optimization 26,
https://doi.org/10.1007/978-3-030-88315-7_2

review of project portfolio optimization problems and approaches to solving such problems using evolutionary and memetic algorithms. In this context, discussions are focused on the problem definitions, mathematical models, and solution approaches from different publications in the literature. It is noted that some of the aforementioned practical steps, such as problem identification and implementation, are outside the scope of this review.

To identify suitable articles for this review, a search of the Scopus[1] database was conducted for articles containing "Project Portfolio", "Project Portfolio + Optimization", and "Project Portfolio + Optimization + Evolutionary" in the title, keywords, or abstract. Note that the search automatically accounts for both the American and British spellings of optimization/optimisation. As of 24 March 2021, the searches returned 8461, 898, and 43 documents, respectively. From the 43 papers returned via the query "Project Portfolio + Optimization + Evolutionary", manual filtering identified 31 suitable articles for this review. The 12 articles not included either did not have full text available or were deemed to be irrelevant.

Additionally, other relevant articles from our earlier work and other sources were also included in this review. It is believed that these articles are adequately representative of the primary focus areas of this book. To the best of the authors' knowledge, articles from predatory publication venues were not included in this review.

From the literature review, the articles on project portfolio optimization can be categorized into the following broad areas:

- Public and Social projects,
- Software/Information Technology (IT),
- Research and Development R&D and Production,
- Construction and Infrastructure,
- Investment, and
- Defense.

The problem context and the definition of the above projects will be discussed in the next section. All these problems were modeled as constrained optimization with either single or multiple objectives, and the developed models were solved using evolutionary algorithms and other approaches. The mathematical models of the problems and approaches for solving the models will be reviewed in the later sections.

2 Problem Description

Most project portfolio problems are defined as project selection problems, such that a set of projects must be selected from a long list of projects. In some cases, the scheduling of the projects is also considered where the start time of each selected project must also be decided. These problems are known as project portfolio selection

---

[1] Available at: https://www.scopus.com/search/form.uri.

and scheduling problems (PPSSPs). This section discusses the various high-level problem descriptions, grouped by application domain, found in the literature.

2.1 *Public and Social Projects*

Projects in the public and social domains are those that receive public funding, generally from a government organization. Quantifying their performance is usually subjective and often based on multiple criteria or objectives. Cruz-Reyes et al. [8] considered multiple groups of independent projects in which each group represents a number of projects from a given geographical region. The central decisions were the selection of projects and a budget allocation process, which were based on the available budget along with an upper and lower limit on the number of projects for each geographical region. Portfolio quality was determined by taking the union of project benefits (social return) across the entire portfolio. Similar problem descriptions can also be found in later works, such as Cruz-Reyes et al. [9–12], Martíez-Vega et al. [31], and Balderas et al. [3]. However, these studies considered different objectives and proposed alternative approaches for decision-making, which will be discussed in a later section. The study of Fernandez and Navarro [16] included priorities among the regions and introduced fuzziness in funding requirements as part of their problem definition. Note that, neither the practical details of the social projects nor their full data were presented in these studies.

Fernandez et al. [15] considered similar regional social projects selection with priority on the selection criteria. Although this included a stationary one-cycle budget, year-wise budget allocation for multiple years is an interesting feature of this problem description. Martínez-Vega et al. [32] described the problem with a multiple-year budget allocation such that a selected project may or may not receive funding in a given year and the available resources may vary from year to year. This time-varying resource allocation problem is termed as dynamic allocation of resources.

Zhang et al. [43] introduced a PPSSP whereby the project selection as well as the project start time must be decided simultaneously. Zhang et al. [43] also considered fuzzy resource constraints to deal with uncertainty. Blecic et al. [4] presented a PPSSP with interdependencies in continuous time. These interdependencies can broadly be categorized as benefit, resource, and technical interdependencies.

2.2 *Software/IT Projects*

Software/IT project selection is a company-level decision making process. Karimpour and Ruhe [26] indicated that inaccuracies due to uncertainty must be considered in defining the software project selection problem. Specifically, they claimed that there was uncertainty pertaining to both the market and the engineering capabilities of the development team. Chiang and Nune [7] considered the selection problem as an optimization problem where an optimal mix of the projects must be determined assuming discrete time-based resource constraints. They also assumed the option of resource divisibility and sharing, task duration variability, time-dependent returns, resource rotation, and learning effect. The project selection as well as the sequencing of the project with their start times are considered by Kremmel et al. [28, 29] and Xiao et al. [41]. In Kremmel et al. [28, 29], project size was determined by the source lines of code measure, with resources formulated as development time, i.e., person-months. The work of Xiao et al. [41] was noted as an extension of [29], with no significant difference in the problem considered. Given that these projects in these studies are scheduled over a planning horizon of multiple periods, they are considered selection and scheduling problems.

2.3 R&D and Production Projects

Projects from the R&D and Production sectors are reported together given that they have significant similarities and overlap in their problem descriptions. George and Farid [18] considered the selection of bio-pharmaceutical, therapeutic antibody projects with activity scheduling. In this study, the timeline was an important aspect associated with project selection. Another important aspect considered in their problem was whether activities were best conducted within the company, outsourced to a contracted researcher, or conducted with a partner company. Kostuik et al. [27] examined offshore exploration and production projects where production is a function of the volume of gas to be injected into each well. Medaglia et al. [33] presented an Research and Development (R&D) project selection problem with full or partial funding options subject to a limited budget. Their model allowed for posterior articulation of preferences and claimed to require the least amount of information from the decision maker. Fernídez et al. [14] considered an R&D project selection problem in manufacturing for new product development that incorporated imperfect knowledge regarding project completion time and budget availability.

2.4 Construction and Infrastructure Projects

Abido and Elazouni [1] and Elazouni and Abido [13] presented multiple problems in the construction domain for coordinated scheduling of activities, whereby the start time for each project must be decided. In their study, the start time of a project is taken as the start time of the first activity. Note that, these problems consist only of the scheduling of activities, i.e., there was no selection aspect.

Iniestra and Gutiérrez [25] introduced a transportation infrastructure project selection problem with multiple selection criteria and a limited budget. This problem is similar to the public project selection problems discussed in Sect. 2.1. Chen et al. [6] described the alternative material supply plans to multiple construction projects, using a transshipment supply chain network, as the contractor's project portfolio. Again, there was no project selection aspect associated with this study.

2.5 Investment Projects

In an investment project selection problem, investors seek to maximize their return by selecting an optimal mix of investment options. In this domain, an investor can be either an individual or a corporation. An important consideration in these types of problems is the risk-reward trade-off, given that investments typically come with some associated risk [30]. Generally, an investor will seek to attain a portfolio that maximizes the expected return while minimizing the risk. In come cases, this can be considered as a multi-objective problem [5]. See [35] for a recent survey on investment stock portfolio selection.

Huang [24] proposed an investment selection problem, assuming a fixed investment amount, with both single- and multi-objective variants. The problem was considered as an abstracted 0/1 knapsack problem. Zhou et al. [45, 46] dealt with a similar problem, but also considered the uncertainty in the portfolio return. Furthermore, multi-term investments were also considered.

2.6 Defense Projects

Although defense projects are publicly funded, they are quite different from social projects in terms of their requirements and prioritization [38]. See [20] for an in-depth review of portfolio optimization in the defense sector. Shafi et al. [37] proposed a defense-oriented PPSSP that encompassed different project groupings that shared a common (central) funding source. This study accounted for many decision making conditions and preferences specific to the defense industry when selecting and scheduling the projects. Xiong et al. [42] described the selection of weapon development projects, similar to the R&D domain discussed in Sect. 2.3. However, Xiong et al. [42] considered both the selection and scheduling of projects, subject to temporal and budgetary constraints. In addition, the variation in performance of a weapon due to the action of counterparts was also considered. Zhou et al. [44] considered a simple weapon selection problem with a limited budget in multiple periods. Harrison et al. [22, 23] proposed a PPSSP inspired by the Future Force Design (FFD) process undertaken by the Australian Department of Defence (DoD). This study examined both the selection and scheduling of a portfolio of projects while adhering to various constraints that were characteristic of the problem faced by the DoD. Harrison et al. [22, 23] then extended their previous work to include grouping into five capability streams, namely Air, Information and Cyber, Land, Maritime, and Space.

2.7 Summary of Problem Descriptions

This section discussed the high-level problem descriptions across six broad application domains. It was observed that the basic structure of the problems was similar, except for the problems in the Construction and Infrastructure domain. Problems can be classified as either 'selection' or 'selection plus scheduling', and either single period or multi-period problems. In all cases, the main consideration for selection and/or scheduling is the available budget. A brief comparison of the problems is presented in Table 1.

3 Problem Formulation

Once a problem description has been analyzed, the next step is the development of a mathematical model. This is an unambiguous mathematical representation of the system, which can be used as a synthetic environment to test alternative decisions without having a direct impact on the real-world system [36]. In this section, the mathematical models developed from the problem descriptions in Sect. 2 are reviewed and their characteristics are discussed.

3.1 Basic Problem Formulation

A basic description of the project portfolio selection problem is as follows. Given a set of n projects, which are competing to be funded by a total budget B, a subset of these projects must be selected by optimizing a given performance measure while adhering to the budget. Each project i has an associated cost, denoted by c_i. In the selection process, a project i can be either selected or not, represented as a binary variable x_i that is set to '1' if the ith project is selected and '0' otherwise. The return (i.e., value) associated with each project i is denoted by r_i. The objective is then to maximize the return through the selection of projects and can be expressed as:

$$\max \sum_{i=1}^{n} r_i x_i. \tag{1a}$$

The budget constraint, which ensures that the total cost must not exceed the available funds, can be formulated as

$$\sum_{i=1}^{n} c_i x_i \leq B \tag{1b}$$

such that the decision variable x_i is binary,

Table 1 Summary of problem descriptions

Problem area	Paper(s)	Problem type	Funding type	Remarks
Social	[3, 8–12, 31]	Selection	Public	Basic selection problem
	[16]	Selection	Public	Fuzziness in budget
	[15]	Selection	Public	Multi-period budget and priority on criteria
	[32]	Selection	Public	Multi-period variable budget
	[43]	Selection and scheduling	Public	Fuzzy resources
	[4]	Selection and scheduling	Public	Interdependencies over time
Software/IT	[26]	Selection	Private	Uncertainty considered
	[7]	Selection	Private	Multi-period resource allocation with resource divisibility and sharing options
	[28, 29, 41]	Selection and scheduling	Private	Sequencing of projects with their start times in multi-period planning horizon
R&D and production	[18]	Selection	Private	Drug selection with activity scheduling
	[27]	Selection	Private	Offshore exploration and production projects
	[33]	Selection	Private	Selection with full or partial funding options
	[14]	Selection	Private	New product development with imperfect knowledge

(continued)

Table 1 (continued)

Problem area	Paper(s)	Problem type	Funding type	Remarks
Construction and infrastructure	[1, 13]	Scheduling	Private	Not a selection problem but multi-project activity scheduling in construction
	[25]	Selection	Public	Transportation infrastructure project with social goal
	[6]	Selection	Private	No selection of projects but material supply plans to multiple construction projects
Investment	[24]	Selection	Private / Personal	Financial/investment options selection
	[45, 46]	Selection	Private / Personal	Financial/investment options selection with uncertainty in return
Defense	[37]	Selection and scheduling	Public / Organizational	Central decision for group-wise project selection and scheduling
	[42]	Selection and scheduling	Public / Organizational	Weapon development projects with performance variation options
	[44]	Selection and scheduling	Public / Organizational	Weapon selection projects
	[22, 23]	Selection and scheduling	Public/Organizational	Long planning horizons with project costs over time
	[21]	Selection and scheduling	Public / Organizational	Long planning horizons with project costs over time and capability stream budgets

$$x_i \in \{0, 1\}, \forall i. \tag{1c}$$

Note that, if we c_i is taken as the weight of an item i, B is taken as a capacity, and x_i is a (binary) decision regarding whether to select an item, the above formulation is equivalent to the binary knapsack problem. Even in this simplistic form, the problem is known to be NP-hard [34] and, hence, there is no known polynomial-time algorithm that can solve arbitrary instances [17, 34].

In social and defense projects, r_i is usually a subjective measure while in other cases, objective values or functions are used. Examples of subjective measures include social benefits (e.g., convenience and the reduction in travel time when considering a new bridge) and community satisfaction. In terms of objective measures, cost and overall benefit are common choices.

Another problem aspect, commonly arising in the social and defense contexts, is group-wise (or region-wise) project selection. In problems with group-wise selection, projects belong to different groups and limits may be imposed on, for example, the number of projects to be selected from, or the spending associated with, each group [11]. However, group-wise allocations are typically allocated from a fixed common pool. In these problem formulations, the decision variable must also account for the group in which a project belongs and can be defined as:

$$x_{ij} = \begin{cases} 1 & \text{if project } i \text{ from group } j \text{ is selected} \\ 0 & \text{otherwise} \end{cases}. \tag{2}$$

Such formulations also require a different set of parameters and data, given by:

c_{ij}	cost of project i in group j
r_{ij}	return from project i in group j
UB_j	upper limit of budget for group j
LB_j	lower limit of budget for group j
B	total budget
P_j	subset of projects in group j

Given the above definitions, the objective function can be expressed as

$$\max \sum_{i \in P_j} \sum_{j} r_{ij} x_{ij}, \tag{3a}$$

subject to the total budget constraint

$$\sum_{i \in P_j} \sum_{j} c_{ij} x_{ij} \leq B \tag{3b}$$

and the allowable upper and lower limits of budget in each group

$$LB_j \leq \sum_{i \in P_j} c_{ij} x_{ij} \leq UB_j, \forall j. \tag{3c}$$

Binary decision variables are used to ensure non-negativity:

$$x_{ij} \in \{0, 1\}, \forall i \in P_j, \forall j. \tag{3d}$$

As a further extension, multiple objectives can be considered, such that a distinct benefit from each group of projects is to be maximized. For a group-wise objective (i.e., one objective function for each group), (3a) can be reformulated as

$$\max \sum_{i \in P_j} r_{ij} x_{ij}, \forall j. \tag{4}$$

In many cases, the budget is distributed over multiple planning periods (/years). However, the decision to fund a project or not must be made at the beginning of the planning horizon. Furthermore, the total budget, cost, benefit, and group-wise limits may vary in each period. A formulation of such a problem is given here.

Firstly, the data and parameters must include a temporal index t:

c_{ijt}	cost of project i in group j in period t
r_{ijt}	return from project i in group j in period t
UB_{jt}	upper limit of budget for group j in period t
LB_{jt}	lower limit of budget for group j in period t
B_t	total budget in period t

The single objective multi-period model can then be formulated as

The objective function:

$$\max \sum_{i \in P_j} \sum_{j} \sum_{t} r_{ijt} x_{ij}, \tag{5a}$$

subject to the total budget constraint in each period

$$\sum_{i \in P_j} \sum_{j} c_{ijt} x_{ij} \leq B_t, \forall t \tag{5b}$$

and the allowable range for the budget in each group and each period

$$LB_{jt} \leq \sum_{i \in P_j} c_{ijt} x_{ij} \leq UB_{jt}, \forall j, t. \tag{5c}$$

Again, the decision variables are defined as binary to ensure non-negativity.

$$x_{ij} \in \{0, 1\}, \forall i \in P_j, \forall j \tag{5d}$$

To incorporate the scheduling with the selection of projects, one must also consider the start time of each project within a multi-period planning horizon as all projects are not required to start at the same time. Moreover, practical constraints, such as precedence or other relationships, may restrict the start times of projects. Each application domain gives rise to various other conditions and constraints, which are considered and discussed below.

3.2 Public and Social Projects

In social projects, it is common to find multi-objective formulations with group-wise selection corresponding to geographical regions, subject to constraints on both the total and group-wise budgets. In general, the objectives are taken as maximization of the benefits observed in each of the regions. Among them, Cruz-Reyes et al. [8] considered 9 objectives and a total of 100 projects with an assumption that the projects are independent, i.e., no precedence relationship existed between any two projects. Similar problem definitions were found in the following studies: Cruz-Reyes et al. [9] considered 3 and 9 objectives; Cruz-Reyes et al. [10] considered 4 and 9 objectives with 25, 100, and 150 projects as well as 9 and 16 objectives with 100, 150, and 500 projects; Cruz-Reyes et al. [12] used 4 objectives with 25 projects and 9 objectives with 100 projects; Cruz-Reyes et al. [11] considered 3 and 8 objectives; and Martínez-Vega et al. [31] used 3, 9, and 16 objectives with 100 projects to solve the corresponding models.

Balderas et al. [3] considered social benefit based on the social category (e.g., extreme poverty, poverty, middle) and the benefit level received (e.g., high impact, middle impact, low impact) for each project as the optimization objectives. Fernandez and Navarro [16] proposed a model with multiple criteria by setting priorities among the regions and introduced fuzziness in the funding requirements constraint.

Fernandez et al. [15] modeled a similar bi-objective regional social project selection problem by maximizing a measure of group satisfaction and minimizing regret. This formulation included year-wise budget allocation and considered up to 500 projects. Using the above model, Martínez-Vega et al. [32] introduced time-varying resource availability in multiple years with the condition that a selected project may or may not receive funding in a given year. Their study considered 20 and 100 projects with 5-, 10-, and 20-year planning horizons.

Zhang et al. [43] formulated a PPSSP with two objectives (maximization of benefit and minimization of risk) such that the project start time was also a decision variable. The constraints included fuzzy resource requirements, synergies between projects, start time windows, precedence relationships, and simultaneous/parallel execution of projects. Blecic et al. [4] developed a single-objective model for PPSSP that aimed to maximize a measure of total performance and included constraints on the budget and interdependencies. This model considered time as continuous and examined a set of 10 projects.

3.3 Software/IT Projects

Karimpour and Ruhe [26] developed a multi-objective model (profit maximization and stability i.e., feasibility) that dealt with uncertainty either as a constraint or an additional objective (for uncertainty related risk) and a constraint on product allocation to segments. Chiang and Nune [7] formulated their problem with both single-objective and multi-objective models. For the single-objective formulation, three objectives (financial benefit, strategic alignment, and portfolio balancing) were combined using a convex combination. For the multi-objective formulation, two objectives were considered (maximizing both financial benefit and strategic alignment). The considered constraints included the budget, risk tolerance, cross-project synergy, and discrete time-based resource availability. They considered an 8-project problem for their study.

Kremmel et al. [28, 29] proposed a five-objective model (revenue, strategic alignment, resource usage, risk, and synergy) with a (discrete) temporal budget constraint. However, it is unclear why some of the objectives were formulated as objectives rather than constraints. The start time of the project was considered as a decision variable, thereby incorporating project scheduling. This study considered a set of 50 projects. Xiao et al. [41] formulated the problem with 2, 3, and 4 objectives (revenue, strategic alignment, resource usage, and risk) with constraints on the start time, required projects, logical project relationships, and project synergies. Note that, the four objectives in the study of Xiao et al. [41] are similar to the first four of Kremmel et al. [28, 29]. This study also considered a set of 50 projects.

3.4 R&D and Production Projects

George and Farid [18] proposed a bi-objective model (maximization of profitability and maximization of the probability of being profitable) with project dependencies. There was no strict budget constraint in this study, but an emphasis was placed on minimizing contracting fees, premiums, royalty charges, and losses in sales revenue to third party entities. Kostuik et al. [27] considered two objectives (maximizing net present value and minimizing capital expenditure) subject to constraints on gas ratio, production, and a profitability index. Medaglia et al. [33] formulated the problem with two objectives (minimizing time-to-market and maximizing economic return) with project interdependency and resource usage constraints. This study also considered full or partial funding options for the projects along with the inclusion of uncertainty via random coefficients in the objective functions. Fernı́ndez et al. [14] proposed a mathematical model with two objectives using imperfect knowledge and a budget constraint. It was assumed that there was no precedence relationship among the projects.

3.5 Construction and Infrastructure Projects

Abido and Elazouni [1] presented a mathematical model with 2 and 3 objectives (cost, duration of the group of projects, and the required credit) subject to constraints on the budget, dependencies between activities (notably not at the project level), project start times, and contractual terms. This study considered two concurrent projects, the first of which had 25 and 30 activities whereas the second had 225 and 240 activities. Elazouni and Abido [13] reformulated this problem with 2 and 3 objectives such that each objective represented the profit from one project, i.e., the two project problem led to a bi-objective formulation while the three project problem led to a formulation with three objectives. The constraints and first instance of the problem were the same as in [1]. For the second instance, projects with 225, 240, and 260 activities were considered.

Iniestra and Gutiérrez [25] developed a model with five objectives (time savings, accident reduction, environmental impact, operational and construction costs, and regional economic development) with constraints on the budget, resource limits in each region, precedence relationships, and mutual exclusivity. The considered problem involved 74 projects in 9 different states. The model proposed by Chen et al. [6] was a single-objective network optimization problem where the objective was to minimize the total supply and material handling costs, subject to flow balance constraints. The considered problem consisted of 7 independent building projects.

3.6 Investment Projects

The mathematical model developed by Zhou et al. [45] consisted of 2 objectives (maximizing return and minimizing risk) subject to constraints on term-wise investment amounts. The actual problem involved 20 projects over 5 periods. However, a subsequent study [46] reformulated the problem with both one-term and multi-term investment options, while a different objective function based on mean semi-covariance theory was employed. Huang [24] considered both single-objective (maximizing return) and bi-objective (maximizing return and minimizing risk) formulations subject to a constraint on the investment amount. In this study, 30 test problems were considered with up to 90 projects.

3.7 Defense Projects

Shafi et al. [37] developed a mathematical model for a group-wise project selection problem with two objectives (minimizing both costs and strategic risk) subject to a budget constraint in each period. An interesting aspect of this model was that it permitted the violation of the budget constraint up to a certain threshold. This study considered 5 groups of projects over a 10 year planning horizon. It was assumed that new projects would be added in each time period—60 projects were considered in the first planning period, with 10 projects being added in each subsequent year such that there were 150 projects available in the final planning year.

The model proposed by Xiong et al. [42] had two objectives (maximizing net present value and effectiveness) subject to constraints on the yearly budget, start and end time bounds, and upper and lower limits on the numbers of weapons of each type. This study considered a hypothetical case with 20 weapon types over a 10 year planning horizon.

Zhou et al. [44] formulated a multi-period, single-objective goal programming model with a limited budget in each period for matching capability. Note that the objective in goal programming is to minimize the sum of all deviations from the set goals.

Harrison et al. [22, 23] developed a single-objective model (maximization of total portfolio value) subject to yearly budget constraints, precedence relationships, and mutual exclusions. This study considered a set of 1000 projects over a 20 year planning horizon and examined various degrees of constraints. A subsequent study by Harrison et al. [21] extended this model to include 5 capability streams, each of which was allocated a fixed proportion of the total budget. Further, this study capped the spending in each year on project initiation and maintenance at 25% and 75%, respectively.

3.8 Summary of Formulations

From the review of mathematical models, it is clear that multi-objective problem formulations are very common when addressing project portfolio selection and/or scheduling problems. However, in many cases, a proper justification for the use of multiple objectives was not provided. In nearly all considered studies, budgetary constraints were necessary. Notably, there was one case involving bio-pharmaceutical development where timing was regarded as more important than budget such that no limit was imposed on project cost. In problems that incorporate a scheduling aspect, it is necessary to include project start times as a decision variable given that they inherently involve multi-period planning. However, some selection problems can also be formulated in terms of multi-period planning such that the budget is distributed annually. Although it is common for precedence relationships to arise as constraints, some of the examined studies made an assumption that all projects were independent. A brief summary of the problem formulations is provided in Table 2.

4 Solution Approaches

It is very common that project portfolio problems are first formulated as mathematical programming models, then solved using optimization techniques. As discussed in the earlier sections, depending on the problem environments, the developed models can be of different types and each model can be solved by using different optimization approaches. Although conventional optimization techniques were widely used in the past, evolutionary computation approaches are currently a popular choice in this field for various reasons. A review of evolutionary computation for solving general optimization problems can be found in Chap. 3 in this book. In this section, a brief review o f he approaches used for the models discussed in Sect. 3 is provided.

Table 2 Summary of problem formulations

Problem area	Paper(s)	Number of objectives	Objectives details	Constraints	Remarks
Social	[8]	9	Group-wise benefits	Budget, group budgets	No precedence relationship
	[9]	3 and 9			
	[10]	4 and 9			
	[12]	4			
	[11]	3 and 8			
	[31]	3, 9, and 16			
	[15]	2			
	[3]	9	Benefits	Same as [8]	Variability in benefits and resources considered
	[16]	1			Priorities among the regions and fuzziness in funding
	[32]	9 and 16			Time-varying resource availability
	[43]	2	Benefits and risk	Fuzzy resource, synergies, start time, precedence relationship, and simultaneous execution	Project start time is a prime decision
	[4]	1	Total performance measure	Budget and inter-dependencies	Interdependencies are in continuous time
Software/IT	[26]	2	Profit and stability	Product allocation to segments	Uncertainty either as a constraint or an additional objective
	[7]	1 and 2	Financial benefit, strategic alignment and portfolio balancing	Budget, risk limit, cross-project synergy, and discrete time-based resource	The single objective is a convex combination of three objectives; first two for two objectives

(continued)

Table 2 (continued)

Problem area	Paper(s)	Number of objectives	Objectives details	Constraints	Remarks
	[28, 29]	5	Revenue, strategic alignment, resource usage, risk and synergy	Period-wise budget	Start time of the project is a decision variable
	[41]	2, 3, and 4	Revenue, strategic alignment, resource usage, and risk	Start time, must select portfolio, logical relationship and synergy	The objectives are similar to the first four of [28, 29]
R&D and production	[18]	2	Maximizing profitability and probability of being profitable	Dependencies	No budget constraint but put emphasis on minimizing fees, premiums, royalty charges, and losses in sales revenue
	[27]	2	Max net present value and min capital expenditure	Gas ratio, production and profitability index	Constraint limits are used to generate sampled portfolio
	[33]	2	Min time-to-market and max economic return	Project interdependencies and resource usage	Full or partial funding options and uncertain elements are included as random coefficients in the objectives
	[14]	2	Imperfect knowledge	Budget	No precedence relationship
Construction and infrastructure	[1]	2 and 3	Cost, duration, and required credit	Budget, dependencies between activities, project start time, and contractual terms	No dependencies at the project level
	[13]	2 and 3	Profit from individual project	Same as [1]	2 (/3) projects implies a 2 (/3) objective problem

(continued)

Table 2 (continued)

Problem area	Paper(s)	Number of objectives	Objectives details	Constraints	Remarks
	[25]	5	Time savings, accident reduction, environmental impact, costs, and regional economic development	Budget, resource limit in each region, precedence relationship and mutual exclusiveness	Region-wise planning
	[6]	1	Total supply and material handling costs	Flow balance	A supply chain network problem
Investment	[24]	1 and 2	Max return, max return and min risk	Investment amount	Single-term decision making
	[45]	2	Max return and min risk	Term-wise investment amounts	Term-wise investment
	[46]	1	Based on mean-semi-covariance	Term-wise investment amounts	Single-term and multi-term investment
Defense	[37]	2	Minimizing both costs and strategic risk	Budget in each period	Budget constraint may be violated up to a certain threshold
	[42]	2	Maximizing net present value and effectiveness	Year-wise budget, start and end time bounds, and upper and lower limits on weapon numbers	Weapon selection study
	[44]	1	Minimizing the sum of all deviations from target	Limited budget in multiple periods for matching capability	Goal programming model
	[22, 23]	1	Maximization of total portfolio value	Budget in each period, project precedence/dependencies, mutual exclusion	Long planning horizons
	[21]	1	Maximization of total portfolio value	Budget in each period, separate budget for project initiation and maintenance, capability stream budgets, project precedence/dependencies, mutual exclusion	Long planning horizons, project selection by capability streams, separate budget for starting and maintaining projects

4.1 Public and Social Projects

Most public and social project formulations utilize a multi-objective optimization model, which is usually solved using evolutionary algorithms. Cruz-Reyes et al. [8] proposed a variant of Non-dominated Sorting Genetic Algorithm II (NSGA-II), referred to as Hybrid Multi-criteria Sorting Genetic Algorithm (H-MCSGA) for solving their 9 objective optimization model. In H-MCSGA, the preference from a Decision Maker (DM) is captured for the most preferred non-dominated solutions. Based on the examined synthetic test problems, H-MCSGA outperformed NSGA-II. However, more experimentation is suggested for small-scale problems to generalize the conclusions. Subsequent work by Cruz-Reyes et al. [9] solved the same model, with 3 and 9 objectives, using NSGA-II and Non-dominated Sorting Genetic Algorithm III (NSGA-III). The main purpose was to compare the performance of these two algorithms in the context of project portfolio optimization. It was concluded that NSGA-III performs better for many objectives than NSGA-II. Note that models with more than 3 objectives are recognized as many-objective optimization.

An additional study by Cruz-Reyes et al. [10] solved the same model again, with 4, 9, and 16 objectives in this instance, using H-MCSGA. In the first phase of H-MCSGA, Non-Outranked Ant Colony Optimization (NOACO) was used to generate a small set of solutions that are classified in (ordered) categories by the DM. In the next phase, a multi-criteria sorting method was combined with an evolutionary algorithm to put a selection pressure towards the Region of Interests (RoI) instead of the whole Pareto front. As such, the DM is directly involved in the optimization phase. Cruz-Reyes et al. [12] then addressed the model, this time with 4 and 9 objectives, using Integrated Preference Disaggregation Analysis (PDA) employing two Multi-objective Evolutionary Algorithms (MOEAs) with differing diversity preservation schemes. The first MOEA was H-MCSGA based on crowding distance (H-MCSGA-CD) and the second was H-MCSGA based on reference points (H-MCSGA-RP). Note that, PDA is an indirect method for preference elicitation. Cruz-Reyes et al. [11] then solved the model, using 3 and 8 objectives, with Multi-objective Evolutionary Algorithm Based on Decomposition (MOEA/D) and MOEA/D-DE (MOEA/D-based on Differential Evolution) instead of NSGA-II and NOACO in the first phase of the hybrid algorithm. In this study, scalable benchmark problems were used for experimentation. Martínez-Vega et al. [31] solved the same model, with 3, 9, and 16 objectives, using a proposed variant of NSGA-III, referred to as A2-NSGA-III. This variant included an adaptive extension that promoted a better spread of the solutions along the Pareto front. The purpose of this study was to compare the performance of A2-NSGA-III, with NSGA-II on project portfolio problems.

Fernandez et al. [15] solved their model by using an extended Non-Outranked Sorting Genetic Algorithm (NOSGA) that also considered preferences on the criteria. In this study, the experimental data consisted of 500 projects. The same model with time-varying resource availability in multiple years was solved by [32] using different MOEAs, including NSGA-II and Strength Pareto Evolutionary Algorithm (SPEA).

Balderas et al. [3] solved their multi-objective optimization model using a variant of NSGA-II, commonly referred to as Grey-NSGA-II (also known as Interval NSGA-II or I-NSGA-II). In this approach, grey numbers are used to reflect the imprecision (or uncertainty) of a magnitude as intervals of real numbers. However, from the interval of solutions, the final solution was then chosen by the DM. Fernandez and Navarro [16] solved both single- and multi-objective models by using multi-attribute value theory and an evolutionary algorithm.

Zhang et al. [43] solved their bi-objective selection and scheduling model using inverse modelling based on a multi-objective evolutionary algorithm using a Gaussian process. Blecic et al. [4] applied Covariance Matrix Adaptation Evolution Strategy (CMA-ES) for solving their single objective selection and scheduling model. In the algorithm, two variation operators were used, namely intermediate recombination and additive Gaussian mutation, along with two different constraint repair methods.

4.2 Software/IT Projects

Karimpour and Ruhe [26] solved their two-objective model using NSGA-II with custom-purpose crossover and mutation operators as well as an additional step for uncertainty sampling.

Chiang and Nune [7] solved both their single-objective (convex combination of three objectives) and bi-objective models using NSGA-II. It was suggested that the projects that contribute the most to strategic goals were not selected when maximization of expected financial benefit was desired.

Kremmel et al. [28, 29] proposed an approach for solving their five-objective model, known as Prototype Optimization with Evolved iMprovement Steps (mPOEMS), as an iterative optimization approach that evolves an EA. Xiao et al. [41] applied MOEA/D for solving their 2-, 3-, and 4-objective models. The study also performed a comparative analysis with NSGA-II and NSGA-III in terms of coverage, distribution, and stability of solutions.

4.3 R&D and Production Projects

George and Farid [18] considered Bayesian networks and evolutionary computation (Estimation of Distribution Algorithm—EDA) for solving their model. In the representation, the chromosome contains 3 components: selection and order, timing, and corporate relation. EDA operates the optimization process checking the dependency and timeline issues, objective space is clustered using a k-means clustering algorithm, and a Bayesian network is used to deal with uncertainty to generate improved population. To get some insights, the objective space is decomposed into three clusters. Kostuik et al. [27] solved both constrained and unconstrained model using MOEA with an innovative constraint handling technique and MILP (Mixed Integer Linear Programming) approach.

Medaglia et al. [33] applied an NSGA-II-based multi-objective evolutionary framework with a posterior articulation of preferences. The initial population was generated based on the concept of linear programming to make sure the individuals are feasible. The objectives were evaluated subject to random elements and fresh individuals were injected to increase diversity. As of their finding, linear constraints favor dealing with large problems. [14] used I-NSGA-II for solving their two objectives model with imperfect knowledge and a budget constraint. However, there are phases in the approach: transforming time effects into an instance involving just intervals in objectives and constraints, solving instance using I-NSGA-II, and finally choosing the best portfolio by the DM.

4.4 Construction and Infrastructure Projects

Both Abido and Elazouni [1] and Elazouni and Abido [13] solved their models using a modified SPEA. They applied SPEA with logic-preserving crossover and mutation operators, and also a fuzzy-based technique to help contractors to select the best compromise solution. Iniestra and Gutiérrez [25] proposed a multi-objective EA with multi-criteria decision making coupled with DM preferences. NSGA-II was used to produce Pareto solutions, and the knee points on the Pareto frontier were identified. The knee points were analyzed using Multi-Criteria Decision Making (MCDM) methods with DMs preference using ELECTRE (Elimination and Choice Expressing the REality). Knee Identification Procedure was used to assist with identifying the most useful 5% of solutions. Chen et al. [6] solved their single-objective network optimization problem using the "Evolutionary Algorithm" module, a part of the Solver add-on, in Microsoft Excel.

4.5 Investment Projects

The mathematical models developed by Zhou et al. [45, 46] were solved by a multi-objective EA with a greedy repair strategy. To estimate the risk effectively, a new kind of mean-semi-covariance was used rather than the standard variance. Huang [24] solved their model using Strength Pareto Evolutionary Algorithm 2 (SPEA2) with a weighted backpack capacity repair policy.

4.6 Defense Projects

Shafi et al. [37] proposed an approach that combined an EA and Reinforcement Learning with a Monte Carlo simulation. Multi-objective Evolutionary Algorithm Based on Decomposition (MOEA/D) was applied to generate a Pareto frontier in each time period. At the same time, reinforcement learning would search horizontally within the set of non-dominated solutions to minimize the capability risks and costs over the entire planning horizon. A multi-period plan was generated with updating data and was re-examined in each time period on a rolling horizon basis. The model developed by Xiong et al. [42] was solved by NSGA-II with a two-point crossover using an integer-valued vector and considered various reactive policies to deal with dynamic changes. Zhou et al. [44] solved a goal programming model with Differential Evolution (DE) using three new strategies, namely variable clustering-based evolution, randomized parameters, and child-individual based selection. Harrison et al. [22, 23] examined the use of a Genetic Algorithm (GA), with two different discrete representations, and the DE and Biased Random-Key Genetic Algorithm (BRKGA) approaches with a real-valued representation. Harrison et al. [21] proposed a hybrid, multi-population approach that evolved both a DE and BRKGA population simultaneously. Knowledge was shared between the populations by periodically sending the best fit individuals, which replaced the worst-fit individuals from the receiving population.

4.7 Summary of Solution Approaches

Traditionally, problems of this nature were formulated as single-objective optimization models, then solved using conventional optimization techniques. More recently, multi-objective modeling is very common when formulating PPSSPs, with evolutionary computation being identified as the preferred approach for solving those models. Among the solution approaches examined, NSGA-II was the most popular approach. A summary of the approaches for the studies reviewed in this chapter is presented in Table 3.

5 Summary

It is first recognized that the keywords used in the Scopus search resulted in a certain type of studies being considered for this review. These studies mainly addressed problems with multiple objectives, modeling them as multi-objective optimization problems, then solving them by using various well-established MOEAs, or variants thereof. This is unsurprising because evolutionary algorithms are recognized as the most suitable approaches for optimizing multiple objectives simultaneously. Among the approaches found in the studies considered in this review, NSGA-II and NSGA-III were the most popular. For NSGA-II, this is likely due to the availability of its source code and its ease of use/modification.

In some of the considered studies, complex practical aspects, such as qualitative performance measures for public/social projects, uncertainty or imprecision in data, dynamic changes over time, and the involvement of the DM in the decision process

Table 3 Summary of solution approaches

Problem area	Paper(s)	Base approach	Approach details	Remarks
Social	[8]	Variant of NSGA-II	Hybrid Multi-criteria Sorting GA (H-MCSGA)	DMs preference is incorporated
	[9]	NSGA-II and NSGA-III	Direct application of NSGA-II and NSGA-III	The performance of these two algorithms is compared
	[10]	H-MCSGA	NOACO with DMs judgment, and multi-criteria sorting	H-MCSGA is implemented in two steps
	[12]	MOEA	H-MCSGA-CD and H-MCSGA-RP	H-MCSGA with two different diversity preservation schemes
	[11]	H-MCSGA	MOEA/D or MOEA/D-DE within H-MCSGA	MOEA/D or MOEA/D-DE in the first phase of the hybrid algorithm
	[31]	Variant of NSGA-III	A2-NSGA-III	NSGA-III with an adaptive extension for better spreading
	[15]	NOSGA	NOSGA with criteria preferences	Preferences on criteria are provided by the DM
	[32]	Different MOEAs	Four MOEAs including NSGA-II and SPEA	All four algorithms were compared for a set of problems
	[3]	Variant of NSGA-II	Grey-NSGA-II /I-NSGA-II	The method deals with imprecision/ uncertainty in data
	[16]	MOEA	Multi-attribute value theory and EA	Fuzzy set is used to model imprecision
	[43]	MOEA	Inverse modeling based on MOEA using a Gaussian process	Fuzzy numbers is used to represent uncertainty
	[4]	CMA-ES	CMA-ES with two variation operators and constraint repairing	With intermediate recombination and additive Gaussian mutation, and two different constraint repair methods

(continued)

Table 3 (continued)

Problem area	Paper(s)	Base approach	Approach details	Remarks
Software/IT	[26]	NSGA-II	NSGA-II with new search operators and an additional step	Custom developed crossover and mutation operators, and an additional step for uncertainty sampling
	[7]	NSGA-II	Direct application of NSGA-II	NSGA-II for both single and two-objective models
	[28, 29]	mPOEMS	Iterative approach that evolves EA	Used smaller time-period, i.e., month instead of year
	[41]	MOEA/D	Direct application of MOEA/D	MOEA/D compared with NSGA-II and NSGA-III
R&D and production	[18]	EDA	EDA for optimization and Bayesian network for dealing with uncertainty	Objective space is decomposed into three clusters using k-means clustering algorithm
	[27]	MOEA	MOEA and MILP	A new constraint handling technique is used with MOEA
	[33]	NSGA-II	NSGA-II with posterior articulation of preferences	Initial population is generated with LP and fresh individuals injected to increase diversity
	[14]	Variant of NSGA-II	I-NSGA-II with DMs involvement	Transforming time effect into interval
Construction and infrastructure	[1, 13]	SPEA	SPEA with logic-preserving crossover and mutation	A fuzzy based technique is used to select the best compromise solution among the population
	[25]	NSGA-II	NSGA-II with MCDM and DMs preference	ELECTRE is used for knee points analysis
	[6]	EA	EA on excel solver	Direct application from MS excel solver module

(continued)

Table 3 (continued)

Problem area	Paper(s)	Base approach	Approach details	Remarks
Investment	[24]	SPEA	SPEA2 with weighted backpack capacity repair policy	It is similar to multi-objective 0-1 knapsack problem
	[45, 46]	MOEA	MOEA with greedy repair strategy	Mean-semi-covariance is used for risk estimation
Defense	[37]	MOEA	MOEA/D with Reinforcement Learning	Multi-period planning on a rolling horizon basis
	[42]	NSGA-II	NSGA-II with a two-point crossover and mutation with integer-valued vector	Various reactive policies applied to deal with dynamic changes
	[44]	DE	DE with three new strategies	The solution is for goal programming
	[22, 23]	DE, GA, BRKGA	Permutation and real-valued representations, operators/parameters selected through hyperparameter optimization	Custom heuristics used to seed population
	[21]	Hybrid, multi-population (DE and BRKGA)	Real-valued representation, operators/parameters selected through hyperparameter optimization	Custom heuristics used to seed population, knowledge exchanged between populations

were considered. In these cases, hybrid/memetic algorithms that combined existing evolutionary computation-based approaches with additional techniques were proposed to generate high-quality solutions. It was also found that hardly any real-world data was used in the examined studies. That is, very few of the studies conducted real-world case studies. Rather, most of the studies made use of synthetic project data.

The initial intention of this review was to provide similar, but easily comparable, information from each paper for each of the examined problem aspects (i.e., problem descriptions, formulations, and solution approaches). This intention was slightly revised given that there were many inconsistencies and a lack of clear information in many of the examined studies. In some cases, the problem description was incomplete, and no justification was provided for developing a particular model, objective function, and/or constraints. However, it is recognized that the justification for choosing or developing a particular methodology was appropriately provided in a majority of cases. Despite this, in nearly all examined studies, it was found that insufficient information on data, or its source, was provided to effectively reproduce the results. This is certainly a major issue for future practitioners in the field of PPSSPs.

References

1. Abido, M.A., Elazouni, A.M.: Multiobjective evolutionary finance-based scheduling: entire projects' portfolio. J. Comput. Civ. Eng. **25**(1), 85–97 (2011). https://doi.org/10.1061/(ASCE)CP.1943-5487.0000070
2. Archer, N., Ghasemzadeh, F.: Project portfolio selection and management. In: Morris, P.W., Pinto, J.K. (eds.) The Wiley Guide to Project, Program and Portfolio Management, pp. 94–113. Wiley, Hoboken (2007)
3. Balderas, F., Fernandez, E., Cruz-reyes, L., Rangel-valdez, N.: Metaheuristic robust optimization of project portfolios using an interval-based model of imprecisions. Int. J. Comb. Optim. Probl. Inform. **7**(3), 101–118 (2016)
4. Blecic, I., Cecchini, A., Trunfio, G.A.: Selection and scheduling problem in continuous time with pairwise-interdependencies. In: Lecture Notes in Computer Science (including subseries Lecture Notes in Artificial Intelligence and Lecture Notes in Bioinformatics), vol. 7334 LNCS, pp. 481–491. Springer, Berlin (2012). https://doi.org/10.1007/978-3-642-31075-1_36
5. Chen, C., Wei, Y.: Robust multiobjective portfolio optimization: a set order relations approach. J. Comb. Optim. **38**(1), 21–49 (2019). https://doi.org/10.1007/s10878-018-0364-9
6. Chen, Q., Garcia de Soto, B., Adey, B.: Transshipment approach to coordinate materials for a contractor's project portfolio. Int. J. Constr. Manag., pp. 1–12 (2020). https://doi.org/10.1080/15623599.2020.1840011
7. Chiang, I.R., Nunez, M.A.: Strategic alignment and value maximization for IT project portfolios. Inf. Technol. Manag. **14**(2), 143–157 (2013). https://doi.org/10.1007/s10799-012-0126-9
8. Cruz-Reyes, L., Fernandez, E., Gomez, C., Sanchez, P.: Preference incorporation into evolutionary multiobjective optimization using a multi-criteria evaluation method. In: Castillo, O., Melin, P., Pedrycz, W., Kacprzyk, J. (eds.) Studies in Computational Intelligence, vol. 547, pp. 533–542. Springer International Publishing, Cham (2014). https://doi.org/10.1007/978-3-319-05170-3_37
9. Cruz-Reyes, L., Fernandez, E., Gomez, C., Sanchez, P., Castilla, G., Martinez, D.: Verifying the effectiveness of an evolutionary approach in solving many-objective optimization problems. In: Melin, P., Castillo, O., Kacprzyk, J. (eds.) Studies in Computational Intelligence, vol. 601, pp. 455–464. Springer International Publishing, Cham (2015). https://doi.org/10.1007/978-3-319-17747-2_34
10. Cruz-Reyes, L., Fernandez, E., Sanchez, P., Coello Coello, C.A., Gomez, C.: Incorporation of implicit decision-maker preferences in multi-objective evolutionary optimization using a multi-criteria classification method. Appl. Soft Comput. **50**, 48–57 (2017). https://doi.org/10.1016/j.asoc.2016.10.037
11. Cruz-Reyes, L., Fernandez, E., Sanchez-Solis, J.P., Coello Coello, C.A., Gomez, C.: Hybrid evolutionary multi-objective optimisation using outranking-based ordinal classification methods. Swarm Evol. Comput. **54**(June 2019), 100652 (2020). https://doi.org/10.1016/j.swevo.2020.100652
12. Cruz-Reyes, L., Perez-Villafuerte, M., Rangel, N., Fernandez, E., Gomez, C., Sanchez-Solis, P.: Performance analysis of an a priori strategy to elicitate and incorporate preferences in

multi-objective optimization evolutionary algorithms. Studies in Computational Intelligence, vol. 749, pp. 401–414. Springer International Publishing (2018). https://doi.org/10.1007/978-3-319-71008-2_29
13. Elazouni, A., Abido, M.: Multiobjective evolutionary finance-based scheduling: Individual projects within a portfolio. Autom. Constr. **20**(7), 755–766 (2011). https://doi.org/10.1016/j.autcon.2011.03.010
14. Fernández, E., Figueira, J.R., Navarro, J.: An interval extension of the outranking approach and its application to multiple-criteria ordinal classification. Omega **84**, 189–198 (2019). https://doi.org/10.1016/j.omega.2018.05.003
15. Fernandez, E., Lopez, E., Mazcorro, G., Olmedo, R., Coello Coello, C.A.: Application of the non-outranked sorting genetic algorithm to public project portfolio selection. Inform. Sci. **228**, 131–149 (2013). https://doi.org/10.1016/j.ins.2012.11.018
16. Fernandez, E., Navarro, J.: A genetic search for exploiting a fuzzy preference model of portfolio problems with public projects. Ann. Oper. Res. **117**(1–4), 191–213 (2002). https://doi.org/10.1023/A:1021577608097
17. Genova, K., Guliashki, V.: Linear integer programming methods and approaches - a survey. Cybern. Inf. Technol. **11**(1), 3–25 (2011)
18. George, E.D., Farid, S.S.: Stochastic combinatorial optimization approach to biopharmaceutical portfolio management. Ind. Eng. Chem. Res. **47**(22), 8762–8774 (2008). https://doi.org/10.1021/ie8003144
19. Ghasemzadeh, F., Archer, N.P.: Project portfolio selection through decision support. Decis. Support. Syst. **29**(1), 73–88 (2000). https://doi.org/10.1016/S0167-9236(00)00065-8
20. Harrison, K.R., Elsayed, S., Garanovich, I., Weir, T., Galister, M., Boswell, S., Taylor, R., Sarker, R.: Portfolio optimization for defence applications. IEEE Access **8**(1), 60152–60178 (2020). https://doi.org/10.1109/ACCESS.2020.2983141. https://ieeexplore.ieee.org/document/9046777/
21. Harrison, K.R., Elsayed, S., Garanovich, I.L., Weir, T., Galister, M., Boswell, S., Taylor, R., Sarker, R.: A hybrid multi-population approach to the project portfolio selection and scheduling problem for future force design. IEEE Access **9**, 83410–83430 (2021). https://doi.org/10.1109/ACCESS.2021.3086070. https://ieeexplore.ieee.org/document/9446148/
22. Harrison, K.R., Elsayed, S., Weir, T., Garanovich, I.L., Galister, M., Boswell, S., Taylor, R., Sarker, R.: Multi-period project selection and scheduling for defence capability-based planning. In: 2020 IEEE International Conference on Systems, Man, and Cybernetics (SMC), pp. 4044–4050. IEEE (2020). https://doi.org/10.1109/SMC42975.2020.9283334
23. Harrison, K.R., Elsayed, S., Weir, T., Garanovich, I.L., Taylor, R., Sarker, R.: An exploration of meta-heuristic approaches for the project portfolio selection and scheduling problem in a defence context. In: 2020 IEEE Symposium Series on Computational Intelligence (SSCI), pp. 1395–1402. IEEE (2020). https://doi.org/10.1109/SSCI47803.2020.9308608
24. Huang, L.F.: A novel multi-objective evolutionary algorithm for portfolio selection. Appl. Mech. Mater. **347–350**, 3128–3132 (2013). https://doi.org/10.4028/www.scientific.net/AMM.347-350.3128
25. Iniestra, J.G., Gutiérrez, J.G.: Multicriteria decisions on interdependent infrastructure transportation projects using an evolutionary-based framework. Appl. Soft Comput. **9**(2), 512–526 (2009). https://doi.org/10.1016/j.asoc.2008.07.006
26. Karimpour, R., Ruhe, G.: Evolutionary robust optimization for software product line scoping: An explorative study. Comput. Lang., Syst. Struct. **47**, 189–210 (2017). https://doi.org/10.1016/j.cl.2016.07.007. https://linkinghub.elsevier.com/retrieve/pii/S1477842416301063
27. Kostuik, K., Valdez, A.A., Bordovsky, J., Mu, Y., Faig, C.: Evaluation of multi-objective methods for portfolio tradeoff analysis. In: SPE Annual Technical Conference and Exhibition. SPE, New Orleans, Louisiana, USA (2013). https://doi.org/10.2118/166372-MS
28. Kremmel, T., Kubalik, J., Biffl, S.: Multiobjective evolutionary algorithm for software project portfolio optimization. In: Proceedings of the 12th Annual Conference on Genetic and Evolutionary Computation - GECCO '10, p. 1389. ACM Press, New York, USA (2010). https://doi.org/10.1145/1830483.1830738. http://portal.acm.org/citation.cfm?doid=1830483.1830738

29. Kremmel, T., Kubalík, J., Biffl, S.: Software project portfolio optimization with advanced multiobjective evolutionary algorithms. Appl. Soft Comput. **11**(1), 1416–1426 (2011). https://doi.org/10.1016/j.asoc.2010.04.013
30. Markowitz, H.: Portfolio selection. J. Financ. **7**(1), 77–91 (1952). https://doi.org/10.1111/j.1540-6261.1952.tb01525.x
31. Martínez-Vega, D., Sanchez, P., Castilla, G., Fernandez, E., Cruz-Reyes, L., Gomez, C., Martinez, E.: Evaluation of the evolutionary algorithms performance in many-objective optimization problems using quality indicators. Studies in Computational Intelligence, vol. 667, pp. 641–653. Springer International Publishing (2017). https://doi.org/10.1007/978-3-319-47054-2_42
32. Martínez-Vega, D.A., Cruz-Reyes, L., Gomez-Santillan, C., Rangel-Valdez, N., Rivera, G., Santiago, A.: Modeling and project portfolio selection problem enriched with dynamic allocation of resources. Studies in Computational Intelligence, vol. 749, pp. 365–378. Springer International Publishing (2018). https://doi.org/10.1007/978-3-319-71008-2_26
33. Medaglia, A.L., Graves, S.B., Ringuest, J.L.: A multiobjective evolutionary approach for linearly constrained project selection under uncertainty. Eur. J. Oper. Res. **179**(3), 869–894 (2007). https://doi.org/10.1016/j.ejor.2005.03.068
34. Pisinger, D.: Where are the hard knapsack problems? Comput. Oper. Res. **32**(9), 2271–2284 (2005). https://doi.org/10.1016/j.cor.2004.03.002
35. Rather, A.M., Sastry, V.N., Agarwal, A.: Stock market prediction and Portfolio selection models: a survey. OPSEARCH **54**(3), 558–579 (2017). https://doi.org/10.1007/s12597-016-0289-y
36. Sarker, R.A., Newton, C.S.: Optimization Modelling: A Practical Approach, 1st edn. Taylor & Francis Group, Boca Raton (2008)
37. Shafi, K., Elsayed, S., Sarker, R., Ryan, M.: Scenario-based multi-period program optimization for capability-based planning using evolutionary algorithms. Appl. Soft Comput. **56**, 717–729 (2017). https://doi.org/10.1016/j.asoc.2016.07.009
38. Tate, D.M., Thompson, P.M.: Portfolio selection challenges in defense applications. Technical report, Institute for Defense Analyses (2017)
39. Urli, B., Terrien, F.: Project portfolio selection model, a realistic approach. Int. Trans. Oper. Res. **17**(6), 809–826 (2010). https://doi.org/10.1111/j.1475-3995.2010.00762.x
40. Wei, H., Niu, C., Xia, B., Dou, Y., Hu, X.: A refined selection method for project portfolio optimization considering project interactions. Expert. Syst. Appl. **142**, 112952 (2020). https://doi.org/10.1016/j.eswa.2019.112952. https://linkinghub.elsevier.com/retrieve/pii/S0957417419306700
41. Xiao, J., Li, J.J., Hong, X.X., Huang, M.M., Hu, X.M., Tang, Y., Huang, C.Q.: An improved MOEA/D based on reference distance for software project portfolio optimization. Complexity **2018**, 1–16 (2018). https://doi.org/10.1155/2018/3051854
42. Xiong, J., Zhou, Z., Tian, K., Liao, T., Shi, J.: A multi-objective approach for weapon selection and planning problems in dynamic environments. J. Ind. Manag. Optim. **13**(3), 1189–1211 (2017). https://doi.org/10.3934/jimo.2016068
43. Zhang, X., Hipel, K.W., Tan, Y.: Project portfolio selection and scheduling under a fuzzy environment. Memetic Comput. **11**(4), 391–406 (2019). https://doi.org/10.1007/s12293-019-00282-5
44. Zhou, Y., Li, Y.B., Shi, Z.Z., Li, Z.X., Zhang, L.: A variables clustering based differential evolution algorithm to solve multistage goal programming model in defense projects portfolio. Adv. Mater. Res. **1046**, 367–370 (2014). https://doi.org/10.4028/www.scientific.net/AMR.1046.367
45. Zhou, Y., Liu, H., Chen, W.: Multi-objective evolutionary algorithm for multi-project and multi-term portfolio problem. In: 2013 IEEE Conference on Computational Intelligence for Financial Engineering & Economics (CIFEr), pp. 55–59. IEEE (2013). https://doi.org/10.1109/CIFEr.2013.6611697
46. Zhou, Y., Liu, H.L., Chen, W., Li, J.: A novel multi-objective evolutionary algorithm solving portfolio problem. J. Soft. **9**(1), 222–229 (2014). https://doi.org/10.4304/jsw.9.1.222-229. http://ojs.academypublisher.com/index.php/jsw/article/view/10132

An Introduction to Evolutionary and Memetic Algorithms for Parameter Optimization

Ali Ahrari and Daryl Essam

Abstract This chapter provides an introduction to the basic ideas and formulations underlying evolutionary algorithms (EAs) for parameter optimization. It discusses the advantages and disadvantaged of EAs in comparison with classical optimization methods. A detailed review of genetic algorithms, the most well-known class of EAs, is provided alongside a general description of other classes (evolution strategies, evolutionary programming, and differential evolution). A few notable relevant search methods are also mentioned. Finally, memetic algorithms are briefly reviewed as extensions to EAs.

1 Introduction

Evolutionary algorithms (EAs) are a class of stochastic population-based optimization methods that have been inspired by the biological evolution process in nature. In EAs, the population of solutions evolves according to the principles of natural evolution, which are mainly selection, mutation, and recombination. More sophisticated EAs may borrow additional concepts from biology; for example, simulating *ecological niches* parallel convergence to multiple optimal solutions, which is refereed to as *multimodal optimization* [12].

Most existing EAs are descendants of three related streams which were introduced to the scientific community in 1960's, although the origin of research on EAs dates back to 1950's [5]. These streams are genetic algorithm (GA), introduced by John Holland [32], evolution strategies (ESs), developed by Rechenberg [55, 56] and Schwefel [58], and evolutionary programming (EP), introduced by Fogel [25, 26]. Despite the similarities in their concept and in some case, in formulation, these streams were developed independently until early 1990's [5, 6]. After that, the inter-

A. Ahrari (✉) · D. Essam
University of New South Wales, Canberra, ACT, Australia
e-mail: a.ahrari@unsw.edu.au

D. Essam
e-mail: d.essam@unsw.edu.au

K. R. Harrison et al., *Evolutionary and Memetic Computing for Project Portfolio Selection and Scheduling*, Adaptation, Learning, and Optimization 26,
https://doi.org/10.1007/978-3-030-88315-7_3

actions between these streams were gradually established, allowing each stream to borrow some ideas from the developments in another one. Later, differential evolution (DE) [63] was proposed for parameter optimization. Besides, there are two other streams of EAs which are not for parameter optimization [22]:

1. Genetic programming (GP) [40] aims to find the optimal program or model with the best fit for given data.
2. Learning classier systems (LCSs) [61] employ the evolution principles to find an optimum set of rules to optimize the behavior of a system for different inputs.

The field of EA has observed an increasing popularity among researchers and practitioners in the recent decades. EAs have been applied to several classes practical problems such as project management [11], vehicle routing [54], job-shop scheduling [28, 44], data mining [27], portfolio management [45], and engineering design optimization [42]. Besides, a number of commercial multidisciplinary optimization software have emerged and been employed by different industries. A more general case of EAs are memetic algorithms (MAs) [48], which utilize individual learning in addition to natural evolution, usually in the form a local search mechanism to improve the efficiency of EAs for specific types of problems.

The rest of this chapter is organized as follows: Sect. 2 compares the flexibility and robustness of EAs with those of classical single-point search methods. Section 3 discusses the general building blocks of EAs. A brief introduction to four mainstreams of EAs for parameter optimization is provided in Sects. 4, 5, 6, and 7. A brief discussion of other relevant metaheuristics is given in Sect. 8. Section 9 provides a brief overview of MAs. Finally, Sect. 10 summarizes this chapter.

2 Comparison Between EAs and Classical Optimization Methods

EAs have generally been developed for *black-box problems*, for which an explicit mathematical definition of the objective function(s) or constraints is not available. This means that there is no information about the fitness landscape of the problem except for the fitness value of an arbitrary solution. Unlike most classical iterative optimization methods, the search process in EAs involves some randomness since the evolutionary operators are stochastic. Besides, EAs do not have a fixed predefined formulation since the mathematical definitions of their operators may change according to the features of the problem. In fact, as stated by Bäck et al. [5], evolutionary algorithms should be interpreted as a style to solve complex optimization problems instead of as a group of general ready-to-use optimization methods. These characteristics of EAs have remarkable effects of their robustness, efficiency, and possibility of customization, which are briefly discussed here.

2.1 Robustness

There is no single optimization algorithm that works well on all types of optimization problems. For each class of problems, the best method can differ from the best method for another class. Developing a new method specialized for each specific problem demands a lot of effort and extensive knowledge about the problem [7]. There are two alternative approaches [7]:

1. formulation of a simpler variant of the actual problem such that the formulated problem meets the assumption of an existing optimization method.
2. Adapting an existing method to the problem at hand.

The first approach is generally the case with classical optimization methods [7] which rely on specific presumptions on uni-modality, continuity, and/or differentiability of the problem landscape. The simplified problem should still capture the main features of the actual problem; otherwise the optimal solution from the formulated problem might be far from the optimum of the actual one. Identification and simulation of all the important features is not a trivial task and requires a good understanding of the problem and the existing optimization methods.

In contrast, EAs are suitable methods for the second approach, thanks to their robustness, flexibility of formulation, and minimal required assumptions. These characteristics enable EAs to deal with complex features that may exist in practical problems, such as:

1. Decision variables of different types: Mixed-variable optimization problems involve different types of variables, e.g. continuous variables, discrete variables, and/or nominal variables [53]. For example, layout optimization of skeletal structures involves a combination of Boolean, discrete, and continuous variables [1].
2. Time-varying landscapes: These problems have one or more features that change over time. Therefore, the global optimum changes over time. If the changes are not radical, an efficient optimization method should be able to track the optimum over time. This type of optimization is known as *dynamic optimization* [49].
3. Hard-to-formulate features: As discussed earlier, optimization methods aim to solve a formulated problem, which, more or less, differs from the actual problem. In particular, there could be some factors which are hard to formulate, e.g., a product's aesthetics, or preference to specific suppliers. If this problem has several (near-) global optima, finding all/most of them can provide the decision-maker with a variety of alternative solutions which may compensate for the missing features of the actual problem. Finding multiple best solutions is known as *multimodal optimization* [12].
4. Conflicting objectives: Many practical problems are subject to conflicting objectives. For example, for a vehicle frame design, its weight should be minimized while crash-safety should be maximized. *multi-objective optimization* [15] can provide a set of solutions illustrating the trade-off between a problem's objectives. Population-based methods are especially useful for this purpose since the final population may provide a good approximation of the best trade-off solutions.

5. Computationally expensive problems: In some real-world problems, especially in engineering design problems, the evaluation of a solution is costly (e.g., when solution evaluation requires an experiment) or time consuming (e.g., when computer simulation is required). The evaluation budget for these problems is thus limited. Surrogate-assisted optimization methods [34] use a surrogate model to approximate a solution value prior to evaluation to make the most out of this limited budget.
6. Uncertain problems: Many problems involve uncertainties in the decision parameters, working conditions, constraints, objective functions, input data, and measurements. *Robust optimization* [8] and *reliability-based optimization* [67] aims to deal with these problems when there are uncertainties in the objective functions and constraints, respectively.

It is notable that a problem may have multiple of these features. For example, *multimodal multi-objective optimization* [66] aims to find multiple of the (near-) global efficient sets of the problem.

Another important advantage of EAs is their capability to perform global search. Even for a simple (but not an easy) problem, the problem landscape may have many local optima which can easily trap most classical optimization methods. The global search capability of EAs help them to avoid such undesirable local optima and converge to the global one, or at least a better local optimum than what single-point local-search based methods can find.

2.2 *Efficiency*

The minimal required assumptions on its problem formulation makes EAs robust; however, this robustness comes at the cost of low efficiency. This means EAs require many more function evaluations (FEs) than classical search methods to converge. The required FEs grows fast with problem dimensionality (faster than linear), resulting in a practical limit on the problem dimension to which they can be applied. For example, in the BBOB'2009 session on global optimization [30], the highest tested problem dimension was 40, and the maximum evaluation budget was set to 10^6D. In contrast, many classical single-point optimization methods can deal with problems with tens of thousands decision variables.

One effective approach to improve the efficiency of EAs is customization. Customization of an EA to a specific class of problems will generally, and in many cases spectacularly, improve the performance of that EA on that specific class, even though the customized method will no longer be applicable to other classes of problems. The customization process involves utilization of some knowledge about the problem on hand to tailor an EAs to it. Such knowledge-based EAs may still enjoy the robustness of EAs to some extent while avoiding unnecessary exploration of the search space, resulting in efficient search methods with some global search capability. A remarkable example is the knowledge-based GA formulated for a specific integer linear programming problem [18, 19], which can deal with the relevant types of problems with up to one billion variables, with an almost linear time complexity (in terms of the required function evaluations).

3 Building Blocks of EAs

EAs generally share the same building blocks. The formulation of these building blocks and the idea behind them may vary from one EA to another and from one problem class to another. These building blocks are:

1. *initialization*: This operator creates a set of (quasi-) random and uniformly distributed solutions in the search space.
2. *selection*: This operator selects two or more parents for generating one or more offspring. Solutions with better function values should have a higher chance for selection.
3. *crossover*: This operator recombines the selected parents to generate one or more offspring. The generated offspring inherits partial attributes of the selected parents. If good attributes of the parents have been inherited, the offspring is likely to be a better solution than both of its parents.
4. *mutation*: This operator performs some random variation to the offspring. Unlike crossover, mutation is applied to each solution independently.

An upper-level description of a typical EA is shown in Fig. 1. After initialization, evolutionary operators (selection, crossover, mutation) are performed iteratively until a termination criterion is satisfied. It is noteworthy that the formulation of these building blocks depend on the employed representation for decision parameters. The most well-known representations are as follows:

1. *Binary Representation*: This type of representation is mostly associated with early GAs, in which the decision variables are coded as binary values. Each solution is thus formed by choosing these binary values. Figure 2 illustrates a binary chromosome representing an exemplary solution for a problem with four decision parameters. Note that this type of representation can only code discrete values; therefore, even if the parameters are naturally continuous, they should be discretized first. Besides, the length of the chromosome depends on the selected precision, this means a higher target precision increases the length of the chromosome and thus a problem's complexity.
2. *Real-valued Representation*: In real-valued representation, there is no need to code its variables as the evolution process is applied to the actual values of decision parameters. This representation can be easily used with continuous decision parameters. It can be used for discrete variables as well, where the value of each variable can be selected from a predefined set of values.
3. *Permutation-based Representation*: Permutation-based representation deals with problems in which a sequence of integer numbers (usually from 1 to D) should

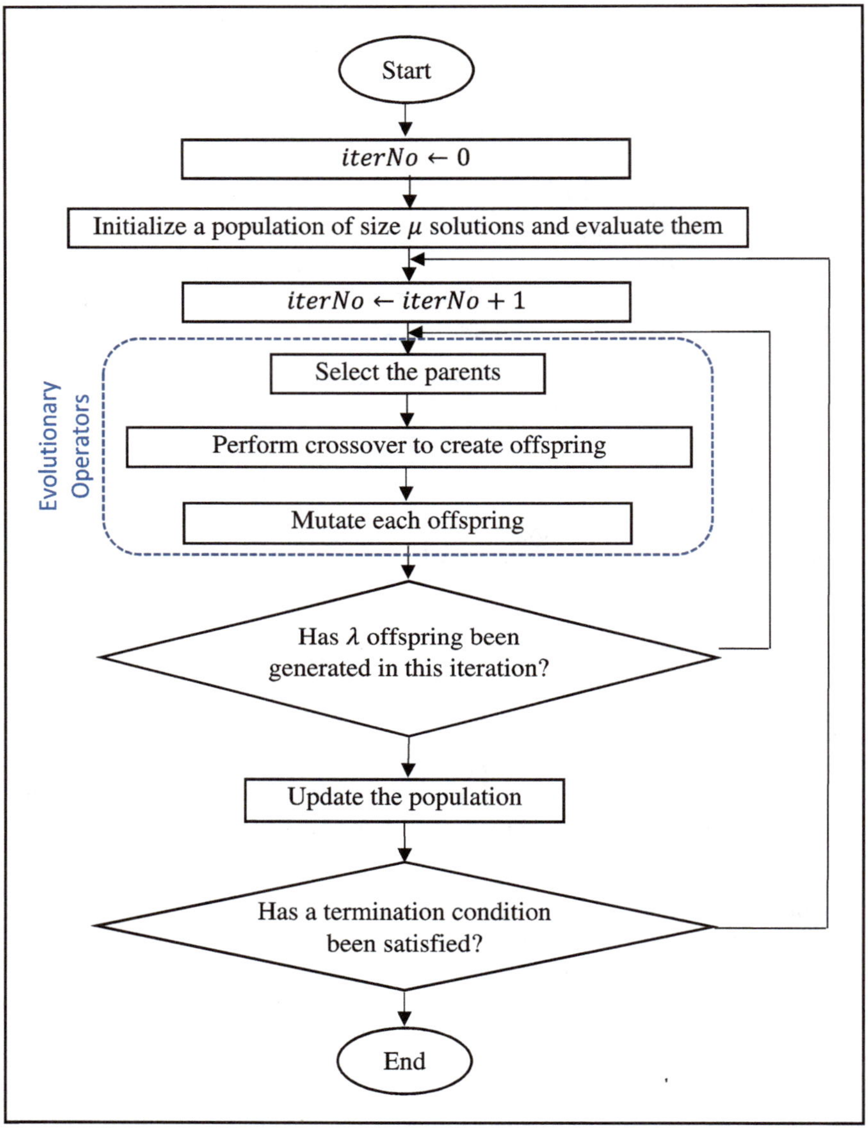

Fig. 1 Flowchart of a simple GA

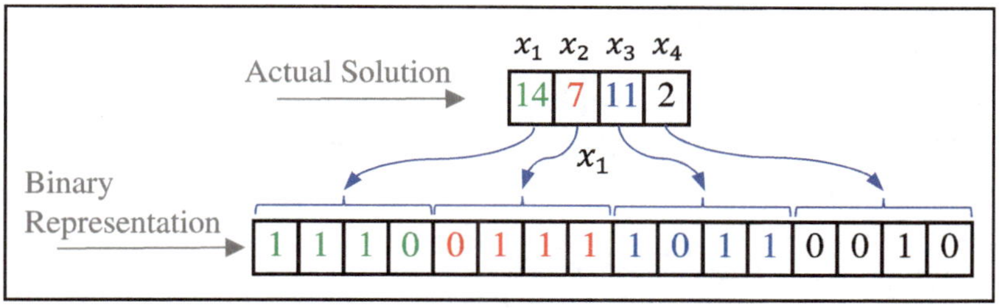

Fig. 2 Binary representation for an exemplary solution for a four dimensional problem

be optimized. This type of representation is mostly encountered in *combinatorial optimization problems* [9] such as traveling salesman problem (TSP) [29]. This type of representation means that each integer value must be in the solution but not be repeated. A solution for this type of representation can be regarded as an ordered list.

4 Genetic Algorithm

Genetic algorithms form the most well-known group of EAs. The building blocks of GAs are further explained in this section.

4.1 Initialization

The purpose of initialization is to create the initial population with maximal diversity to maximize exploration in the early iterations. Quite often, a random generation of solutions in the search space is sufficient for this purpose. However, for computational expensive problems with very limited evaluation budgets, more sophisticated space filling methods may be beneficial [35].

4.2 Selection

The selection process assures that better solutions have a higher chance for reproduction. There are several selection operators, two of the most well-known ones are explained in this section.

4.2.1 Roulette Wheel Selection

The roulette wheel selection assigns a probability of selection to each solution proportional to their fitness. Assuming the problem is a maximization problem, the probability of selecting solution $\boldsymbol{x}_i$ is calculated as follows:

$$p_i = \frac{\hat{f}_i}{\sum_{j=1}^{N} \hat{f}_j} \tag{1}$$

In this equation, $\hat{f}_i$ is the normalized fitness of solution $\boldsymbol{x}_i$. Note that this normalization is often required because:

1. The fitness of a solution might be negative, resulting in a negative value for p_i.
2. The selection pressure will be very strong in the early iterations when the difference between the fitness of solutions is huge. This may results in very strong exploitation in early iterations. In contrast, the selection pressure may become too small when the population is converging, which results in undesirable weak exploitation.

The normalization can be performed as follows:

$$\hat{f}_i = a_0 + \frac{f_i - f_{\min}}{f_{\max} - f_{\min}} \tag{2}$$

in which $f_{\min}$ and $f_{\max}$ are the lowest and highest fitness value in the population and $a_0 > 0$ is a constant controlling the section pressure. The highest to lowest p_i at each iteration is $(1 + a_0)/a_0$.

4.2.2 Tournament Selection

Even with linear scaling, the roulette selection has a major drawback. It is sensitive to strictly monotonic transformations of the objective space. For example, optimization would be substantially different if $\exp(f(x))$ is optimized instead of $f(x)$.

Rank-based selection schemes overcome this sensitivity since the rank of solutions do not change when a strictly monotonic transformation is applied to the objective space. Besides, they do not need normalization of the objective function. In particular, tournament selection has become a popular selection scheme do to its simplicity and robustness.

In tournament selection, first, a number of K solutions are randomly selected and the best one becomes the parent for crossover. This process is repeated to choose the second parent for crossover. Parameter K is the tournament size and it is directly related to the selection pressure. Figure 3 shows the numerically calculated selection probability (p_i) for the ith parent when there are 20 parents sorted according to their fitness. As can be observed, the selection pressure intensifies by increasing K. The popular choice of $K = 2$ results in almost linearly decreasing selection weights.

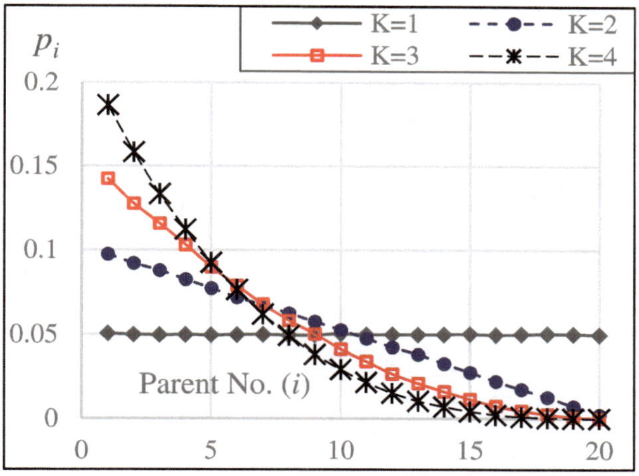

Fig. 3 Effect of the tournament size (K) on the selection probability (p_i) of each parents. There are 20 parents, which are sorted according to their fitness (the first parent is the best one)

4.3 Crossover

There are numerous formulation for crossover for different types of representation. However, the idea underlying all crossover operators is the same: children should inherit the properties of their parents. A few well-known crossover operators are explained in this section.

4.3.1 Crossover for Binary Representation

GAs that use binary representation are known as Binary GAs (BGAs). One-point crossover is the earliest crossover operator for BGAs. For two selected parents p_1 and p_2, this operator selects a random location on the chromosome to divide it into two parts. The children inherit the first part from one parent and the second part from the other parent. In two-point crossover, there are two crossover locations. Each child inherits the genes of the first and the third part from one parent and the middle part from the other parent. Figure 4 illustrates examples of one-point and two-point binary crossovers. This type of crossover can easily be extended to K-point crossover, where K is a natural number. Another type of generalization is the uniform crossover, in which the value of each gene is inherited from either of the parents with equal probability.

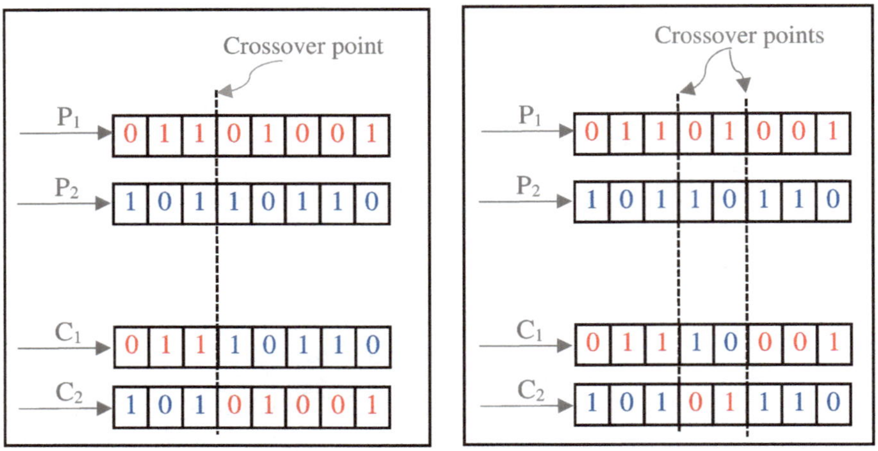

(a) One-point crossover (b) Two-point crossover

Fig. 4 One-point and two point crossover for binary GA

4.3.2 Crossover for Real-Valued Representation

The crossover operator for real-valued GAs demands a different formulation because simply swamping the values of some decision variables provides only small possibilities for new solutions. Therefore, it is rational to employ a probability distribution for continuous variables. There are several well-known crossover operators for Real-valued GAs (see [36] for review) such as Simulated Binary Crossover (SBX) [16] and parent-centric crossover (PCX) [17]. SBX is explained in this section as an example.

SBX intends to simulate the behavior of one-point binary crossover in a continuous search space. Here the case for a one variable problem is discussed first, and then the idea is generalized to D-dimensional problems.

In SBX, the centroids of children and parents coincide. Besides, SBX defines a *spread factor* β which determines the ratio of distance between the children and that of the parents. For $\beta < 1$, children are closer to each other than the parents (contraction) whereas for $\beta > 1$ expansion occurs. These two features can be formulated as follows:

$$\begin{aligned} c_1 + c_2 &= p_1 + p_2 \\ \beta &= \frac{\|c_1 - c_2\|}{\|p_1 - p_2\|}. \end{aligned} \tag{3}$$

The value of β is selected from the following density function:

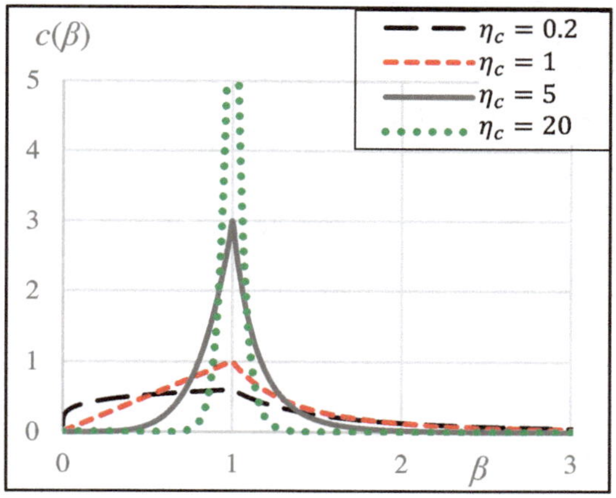

Fig. 5 Effect of the crossover index (η_c) on the density function of β, denoted by $c(\beta)$

$$c(\beta) = \begin{cases} 0.5(\eta_c + 1)\beta^{\eta} & \text{if } \beta \leq 1 \\ \dfrac{0.5(\eta_c + 1)}{\beta^{\eta+2}} & \text{otherwise} \end{cases}, \tag{4}$$

in which $\eta_c > 0$ is the index of SBX. It is notable that this density function is always maximal at $\beta = 1$; however, a higher value of η_c means the sampled values of β are more likely to be closer to one. The effect of η_c on $c(\beta)$ is depicted in Fig. 5. For a given value of β, the children can be easily generated using (3) as follows:

$$\begin{aligned} c_1 &= 0.5[(1 + \beta)p_1 + (1 - \beta)p_2] \\ c_2 &= 0.5[(1 - \beta)p_1 + (1 + \beta)p_2] \end{aligned} \tag{5}$$

This formulation may result in children that are outside the search space if expansion occurs. However, it is possible to improve SBX such that the children are always generated inside the search space. A simple approach is to find the maximum values of $\beta_{\max}$ such that the children are generated inside the search space. There should be two values of $\beta_{\max}$, one for ensuring that the child is less than the upper limit (x_U) and the other to make make sure that it is greater than or equal to the lower limit of the search space x_L.

SBX can easily be generalized to multi-variate problems using a one variable at a time strategy: for each variable, SBX is performed with a probability of P_c. Next, the children swap the values of each pair of variables with a probability of P_{swap}. This swap procedure generates children around a grid of points determined by the parents and the coordinate system. Figure 6 shows the effect of the SBX index and the swap probability on the distribution of 200 randomly generated children for two parents.

As can be observed, SBX with $P_{\text{swap}} > 0$ or $P_c < 1$ noticeably favors search along coordinate directions. This bias can be very beneficial when local minima are along coordinate directions: a case often found in separable problems. How-

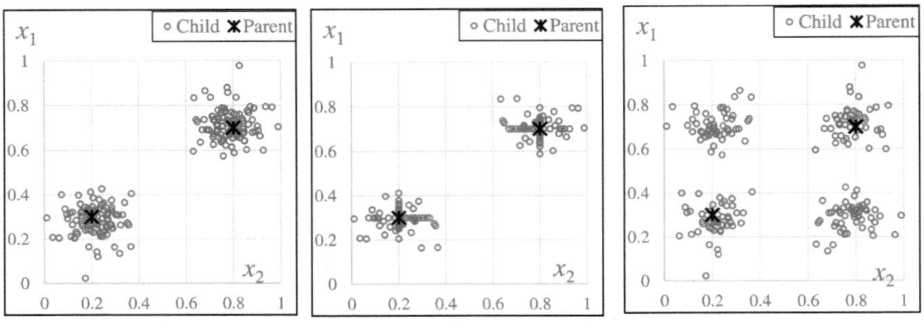

(a) $P_c = 1.0$, $P_{\text{swap}} = 0$ (b) $P_c = 0.5$, $P_{\text{swap}} = 0$ (c) $P_c = 1$, $P_{\text{swap}} = 0.5$

Fig. 6 Distribution of 200 sampled children by crossover of two fixed parents using SBX with different values for P_c and the swap probability (P_{swap}) when $\eta_c = 3$

ever, when this separability is removed, a drastic decline in the performance of this operator, and in general any operator that favors search along coordinate directions, can be observed. Rigid rotation of the search space [57] can eliminate this separability without changing other aspects of the problem landscapes. A comparison of many stochastic population-based search methods, which are generally known as *metaheuristic methods*, in the CEC'2005 test problems for real parameter optimization [64], reveals that the majority of them suffer from dependency on this specific characteristic of the separable problem. Since EAs are expected to be robust search tools, this dependency is a severe drawback. Performing SBX in a randomly selected coordinate system can overcome this dependency. It can improve the performance of SBX in the rotated landscape in exchange for losing outstanding outcomes in the presence of separability. An adaptive mechanism that can learn the separable and non-separable variables is the ideal choice. However, developing such mechanisms is not an easy task.

4.3.3 Crossover for Permutation Representation

There are several crossover operators for permutation representations. In general the sequence of numbers in the child should share some similarity with its parents. There are several crossover operators for this purpose (see [36] for some examples). The order one crossover (OX1) is explained in this section.

OX1 determines a crossover point in the parents, which divides the chromosome into two substrings. The first child directly inherits its first part from the first parent. The second part of this child is determined by the order of the remaining values in the second parent. This process is illustrated in Fig. 7.

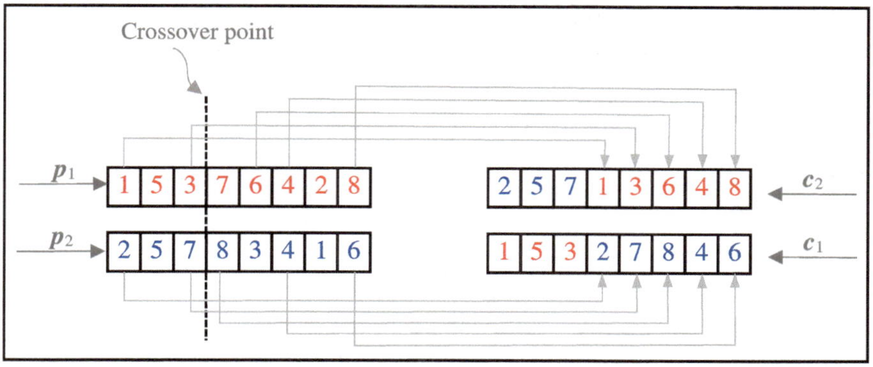

Fig. 7 Order one crossover (OX1) for permutation representation

4.4 *Mutation*

The mutation operator in GAs improves exploration (diversification). It applies some random variation to one solution. This random variation is disruptive, especially in the late iterations of the optimization process when the population has already found a good region of the search space. However, it occasionally results in better and different solutions which can direct the subpopulation to (potentially) a better region of the search space. Like crossover, the formulation of the mutation operator depends on the employed representation, and in some cases, on the problem itself.

4.4.1 Mutation for Binary Variables

The simplest way to mutate a binary chromosome is to flip the value of a bit. For each bit, the value changes (from zero to one or from one to zero) with a probability of P_m. By default, the value of P_m is inversely proportional to the length of the chromosome.

4.4.2 Mutation for Continuous Variables

For continuous values, the mutation applies some random variation following a continuous density function. The strength of this variation is defined by the user, dynamically reduced (for better exploitation in the final stages of optimization), or ideally, (self-) adapted. For example, in polynomial mutation, first, a random number ($r \in [0, 1]$) is generated, then mutation is performed as follows:

$$x^{\text{mut}} = \begin{cases} x + \bar{\delta}_L(x - x_{\min}) & \text{if } r \leq 0.5 \\ x + \bar{\delta}_R(x_{\max} - x) & \text{otherwise} \end{cases}, \tag{6}$$

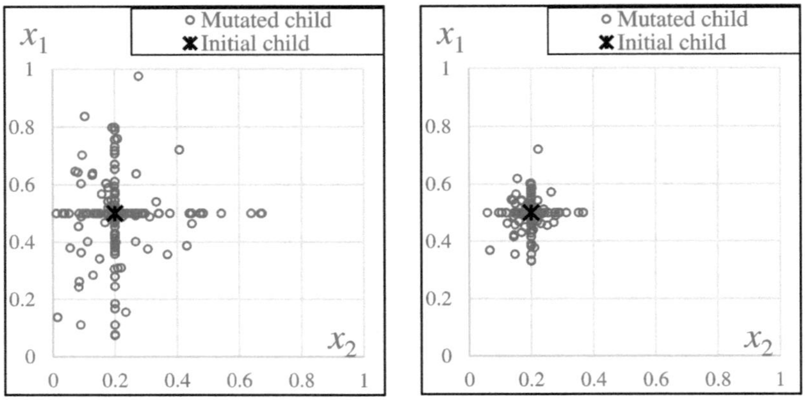

(a) $\eta_m = 5, P_m = 0.5$ (b) $\eta_m = 20, P_m = 0.5$

Fig. 8 Polynomial mutation for selected values of η_m and P_m

in which $-1 \leq \bar{\delta}_L \leq 0$ and $0 < \bar{\delta}_R \leq 1$ defines variation strength relative to the bounds of the search space. For $r < 0.5$ the x^{mut} will fall on the left side of x, whereas for $r > 0.5$ if falls on the right side of x. The density function for these values should be selected such that larger variants are less probable. In polynomial mutation, the density function is defined as follows:

$$
\begin{aligned}
c(\bar{\delta}_L) &\propto \begin{cases} (1+\bar{\delta}_L)^{\eta_m} & \text{if } -1 \leq \bar{\delta}_L \leq 0 \\ 0 & \text{otherwise} \end{cases} \\
c(\bar{\delta}_R) &\propto \begin{cases} (1-\bar{\delta}_R)^{\eta_m} & \text{if } 0 \leq \bar{\delta}_R \leq 1 \\ 0 & \text{otherwise} \end{cases}
\end{aligned}
\tag{7}
$$

$\eta_m > 0$ controls the mutation strength such that a smaller η_m means a stronger mutation. Using these density functions, $\bar{\delta}_R$ and $\bar{\delta}_R$ can be easily calculated:

$$
\begin{cases} \bar{\delta}_L = ((2r)^{1/(\eta_m+1)} - 1 & \text{if } r \leq 0.5 \\ \bar{\delta}_R = 1 - (2(1-r))^{1/(\eta_m+1)} & \text{otherwise} \end{cases}
\tag{8}
$$

For multivariate problems, this mutation is applied to each variable independently with a probability of P_m. Figure 8 depicts 200 solutions generated from the mutation of an initial solution x in two-dimensional space when the search space is $[-1, 1]^D$. As can be observed, a greater η_c means smaller variations. Besides, this type of mutation favors search along coordinate directions, especially when P_m is small.

4.4.3 Mutation for Permutation Representation

The mutation operator for an ordered list, which is a solution in the permutation-based representation, aims to add some random variation to the existing order. Some well-known mutation operators for this purpose are

1. *insertion mutation*: this operator simply relocates one value to a random position in the solution.
2. *inversion*: this operator selects a substring from the solution and inverts the sequence of values in this substring.
3. *shuffle mutation*: This type of mutation selects a substring of the solution and shuffles the values of the variables in this substring.
4. *random swap mutation*: This operator selects two random variables and swaps their values.

The strength of this random variation should be controllable either by the user or by a (self-) adapting mechanism in the algorithm. For example, the insertion mutation may occur with a probability of $P_m = 1/D$ for each variable.

4.5 Population Update

After generating the population of children, the population should be updated for the next generation. This update can be considered as another stage in the selection process, which is often deterministic. At this stage of generation, there are μ parents which have generated λ children. Depending on whether the parents may survive to the next generation, the update mechanism can be categorized as follows:

1. *Non-elitist selection*: μ-best children are selected as the parents for the next generation. This is also known as *comma selection* and is denoted by (μ, λ)-GA.
2. *elitist selection*: μ parents for the next generation are selected from the union of children and $0 \leq N_{\text{elite}} \leq \mu$-best parents.
3. *steady-state selection*: μ-best solutions are selected from the union of current parents and children. This is also known as *plus selection* and denoted by $(\mu + \lambda)$-GA.

The more general case of elitist selection becomes identical to comma and plus selection schemes for $N_{\text{elite}} = 0$ and $N_{\text{elite}} = \mu$, respectively. The choice of non-elite selection, which is the case with the generic genetic algorithm, is counter intuitive as it may lose the best solution found so far. The steady-state selection sounds like the most reasonable choice; however it is prone to be trapped in local optima since one or a few good solutions generated in the early iterations may attract the whole population without giving it sufficient time to explore the search space. The elite selection provides a trade-off between the advantages of the extreme selection schemes.

4.6 Stopping Criteria

The evolutionary search process continues until one stopping criterion is met. The simplest stopping criterion is based on a predefined number of function evaluations or generations. More sophisticated stopping criteria may consider variation of solutions in the decision variable space or objective value space.

5 Evolution Strategies

Similar to GAs, evolution strategies (ESs) follow principles of natural evolution to evolve a subpopulation. In contrast to early GAs which employed binary representation, ESs have been mainly developed to deal with continuous variables. In the general form, at each generation, λ solutions are generated by μ parents. This process involves recombination of ρ parents. The evolutionary operators of ESs are briefly explained in this section.

5.1 Selection

Selection in ESs is a deterministic process. It is usually made over the child population (comma scheme) or the union of children and parents (plus scheme), although more general cases are possible, such as limiting the life span of each solution to $1 \leq \kappa$ generations [5] or gradually reducing the fitness of older solutions [2].

The first evolution strategy was a (1+1)-ES [8], in which one child is generated by mutation of one parent, where the child replaces the parent if it has a better fitness. This simplest structure later evolved to (μ+1)-ES, and subsequently, ($\mu/\rho+\lambda$)-ES and ($\mu/\rho,\lambda$)-ES were developed in which ρ parents recombine to generate a child.

5.2 Recombination

There are two main types of recombination for ESs:

1. *Discrete recombination*, which resembles the uniform crossover in GAs. The child inherits the value for each variable from either of the parents with equal probability (Fig. 9a). This type of recombination is more suitable for discrete variables.
2. *Intermediate recombination*, in which the child is the (weighted) average of the parents (Fig. 9b), which makes it more suitable for continuous variables.

The temporary solution obtained by recombination is called the *recombinant*, which will be mutated to form a new solution.

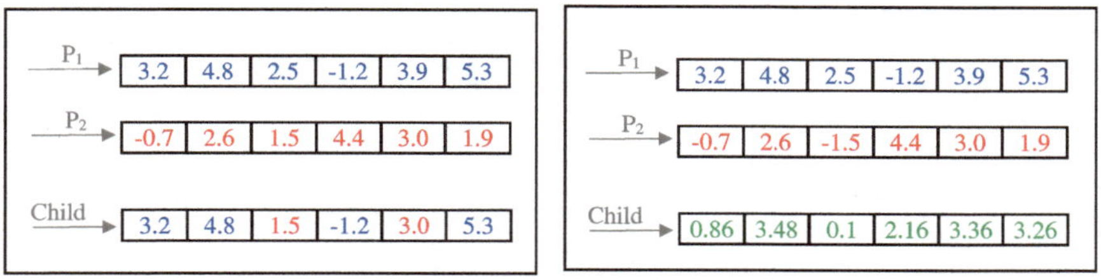

(a) Discrete recombination

(b) Intermediate recombination ($Child = 0.4P_1 + 0.6P_2$)

Fig. 9 Examples for discrete and intermediate recombination for two parents, each having six variables

Both types of recombination can be used with an arbitrary number of parents. The preferred choice in more recent ESs is global (weighted) recombination, a variant of intermediate recombination in which all parents participate in recombination of children, with possibly different weights, to form the recombinant (e.g., the recombination type in covariance matrix adaptation evolution strategy (CMA-ES) [31], known as the state-of-the-art ES). It is remarkable that this type of recombination does not require selecting a new set of parents for generation of each child.

5.3 Mutation

Mutation is the primary operator in ESs. It is generally performed using a normal distribution:

$$\boldsymbol{x}_i = \boldsymbol{x}_{\text{mean}} + \sigma_{\text{mean}} \mathcal{N}_D(\mathbf{0}, \mathbf{C}) \tag{9}$$

In which $\mathcal{N}_D(\mathbf{0}, \mathbf{C})$ is a random number sampled from a multivariate normal distribution with the covariance matrix $\mathbf{C}$. σ_{mean} and $\mathbf{C}$ define the mutation profile. In the simplest scenario, $\mathbf{C}$ is the identity matrix. In this case, the distribution of sampled solutions is isotropic (Fig. 10a), in which the mutation strength, also known as step size, is σ_{mean}. In a more general case, $\mathbf{C}$ is a diagonal matrix in which the distribution of samples forms an axis-parallel ellipsoid (Fig. 10b). In the most general case, $\mathbf{C}$ is a positive definite symmetric matrix, resulting in a rotated ellipsoid distribution of the sampled solutions (Fig. 10c).

The importance of more complex mutation profiles can be observed in Fig. 11 for three simple benchmark functions. For the Sphere function, isotropic mutation is sufficient to efficiently converge to the global optimum. For the elliptic function, using this type of mutation results in a slow progress rate. A diagonal $\mathbf{C}$ can substantially

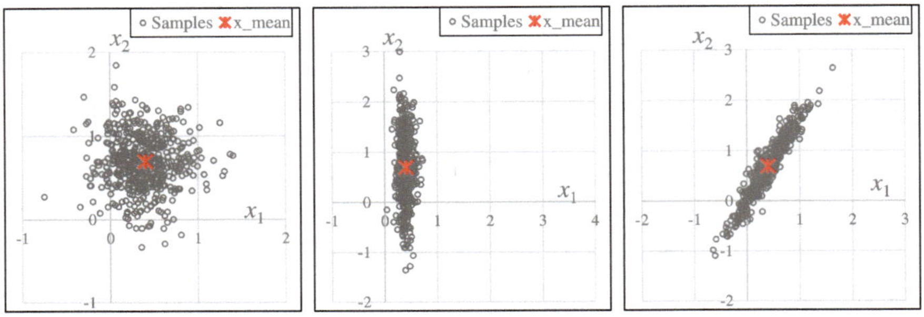

(a) (b) (c)

Fig. 10 Exemplary distribution of 500 sample solutions when **C** is **a** the identity matrix, **b** a diagonal matrix, **c** a general symmetric positive definite matrix

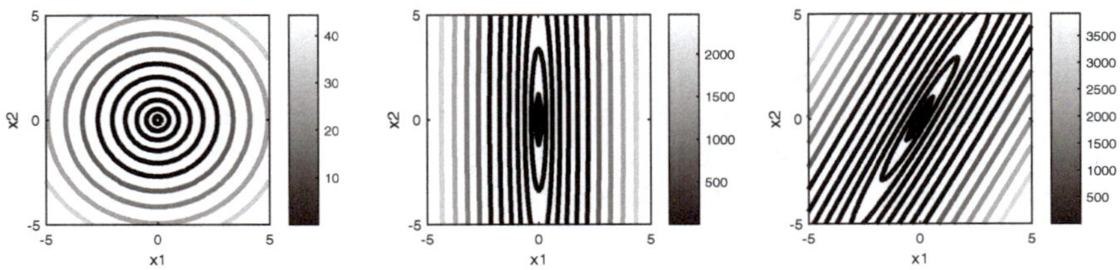

(a) Sphere function (b) Elliptic function (c) Rotated elliptic function

Fig. 11 Contour lines of three simple functions

improve the convergence efficiency. For the rotated elliptic function, adjusting the full-covariance matrix can results in considerable improvement in the convergence rate.

5.4 Adjusting the Mutation Profile

On-line adjustment of the mutation profile is the major issue in ESs. Some of the main developments in adjusting the mutation profile are explained in this section.

5.4.1 One-Fifth Rule

The one-fifth rule is the first adaption mechanism which was developed for (1+1)-ES [59]. It simply calculates the fraction of times that the child is better than its parent. If this number if greater than 1/5, σ_{mean} is increased, otherwise it is decreased to improve exploitation.

5.4.2 Self-adaptation

Self-adaptation is a more general approach that is applicable to population-based ES [8]. It uses a slightly different step size to generate each solution:

$$\sigma_j = \sigma_{\text{mean}} \exp(\tau \mathcal{N}(0, 1)). \tag{10}$$

This slightly altered mutation strength is then used to generate the jth child:

$$\boldsymbol{x}_j = \boldsymbol{x}_{\text{mean}} + \sigma_j \mathcal{N}_D(0, 1)). \tag{11}$$

It is expected that a better σ_j is more likely to result in a better solution. Therefore, after selecting the μ-best solutions as the parents for the next generation, σ_{mean} is set to the (weighted) average of σ_j of the new parents.

Self-adaption can be generalized to the case when the covariance matrix is diagonal and even when a full covariance matrix should be adjusted. However, it has one major drawback: it focuses on immediate success, even if it is insignificant. Smaller variations from the recombinant often results in a better solution; however a larger variation may result in substantial improvement in the long-term. Consequently, self-adaptation generally results in smaller-than-optimal mutation strength, a problem referred to as *opportunism* [8].

5.4.3 Cumulative Step-Size Adaption

Cumulative step-size adaption (CSA) [51] is a deterministic approach to update σ_{mean}. It takes into account the history of the changes in $\boldsymbol{x}_{\text{mean}}$ to calculate the evolution path. This calculation considers not only the size of the change in $\boldsymbol{x}_{\text{mean}}$ but also whether the directions of these changes are correlated. The size of the evolution path update is then compared with the expected size of the evolution path if the parents had been selected randomly. If the former is greater than the latter, σ_{mean} is increased; otherwise it is decreased. CSA does not suffer from opportunism, possibly because it employs historical information (with an exponential decay rate) instead of immediate success of individual solutions.

5.4.4 Adaption of the Covariance Matrix

Hansen and Ostermeier [31] proposed a deterministic approach to update the full covariance matrix. This adaptation method has two main components:

1. *rank μ update*, which accounts for the distribution of the selected parents at each iteration. This component forms the matrix $\mathbf{C}_\mu$ by the search directions and extents suggested by the selected parents. The matrix rank of $\mathbf{C}_\mu$ is μ.

2. *rank one update*, which takes the change in the evolution path of $\boldsymbol{x}_{\text{mean}}$ into account to define $\mathbf{C}_1$. The matrix rank of $\mathbf{C}_1$ is one.

The updated covariance matrix is a weighted average of the current covariance matrix ($\mathbf{C}$), $\mathbf{C}_\mu$, and $\mathbf{C}_1$, plus another minor correction term. Combined with CSA, this adaptation method is a peculiar feature of CMA-ES [31], the state-of the-art evolution strategy for continuous single objective optimization. CMA-ES can effective deal with rotated ill-conditioned problems thanks to its efficient adaptation of the covariance matrix to the fitness landscape. Besides, the search process in CMA-ES is independent of the coordinate system. As a result, the performance of CMA-ES remains unchanged if the landscape is linearly rotated, an interesting feature which does not exist in many metaheuristic search methods. Several variants of CMA-ES have later been developed, usually with minor modifications [4].

6 Evolutionary Programming

Evolutionary programming (EP) was initially developed as a machine learning tool for finite-state machines; however, contemporary EP methods are used for numerical function optimization. EP shares a lot of features with ESs, and the field of EP has almost merged into ESs [22].

In the basic variant of EP, no recombination is performed, and each parent generates one child by mutation. The mutation strength depends on the fitness of the parent. Given the parent $\boldsymbol{x}_j^{\text{old}}$, the mutation strength is calculated as follows:

$$\sigma_i = \left(\beta_i \bar{f}(\boldsymbol{x}_j^{\text{old}}) + \gamma_i\right)^{0.5}, \tag{12}$$

in which $\bar{f}(\boldsymbol{x}_j^{\text{old}}) > 0$ is the scaled fitness function. The child corresponding to this parent is then generated as follows:

$$\boldsymbol{x}_j^{\text{new}} = \boldsymbol{x}_j^{\text{old}} + \sigma_i \mathcal{N}_D(0, 1). \tag{13}$$

After generation of μ children (one for each parent), a stochastic tournament selection is performed to choose μ solutions from the union of existing parents and generated children. These selected solutions then serve as the parents for the next generation.

7 Differential Evolution

Differential Evolution (DE), introduced by Storn and Price [63], forms a more recent stream of EAs for continuous parameter optimization. Like other streams, it relies on recombination, mutation, and selection. In contrast to ESs and GAs, mutation is performed before recombination in DE.

7.1 Mutation

In the generic DE, mutation adds a weighted difference of two solutions to the third one. Let $\boldsymbol{x}_i$ be the selected solution for mutation, which is referred to as the *target solution*. A mutation vector for this target solution is determined as follows:

$$\boldsymbol{v}_i = \boldsymbol{x}_{r_1} + F(\boldsymbol{x}_{r_2} - \boldsymbol{x}_{r_3}), \tag{14}$$

in which i, r_1, r_2 and r_3 are distinct indexes, the constant $0 \leq F \leq 2$ determines the scale of the differential variation, and $\boldsymbol{v}_i$ is the mutant vector for the solution $\boldsymbol{x}_i$.

7.2 Crossover

The crossover operator employs the mutant vector. First, it defines a *trial vector* as follows:

$$u_{ij} = \begin{cases} v_{ij} & \text{if } (rand \leq CR) \text{ or } j = j_0(i) \\ x_{ij} & \text{otherwise} \end{cases}, \; j = 1, 2, \ldots, D. \tag{15}$$

In this equation, *rand* is a uniformly distributed random number in [0, 1] sampled anew for each j and each i, CR is a user-defined control parameter called the *crossover constant*, and $j_0(i)$ is a randomly selected index in $\{1, 2, \ldots, D\}$ so that at least one variable takes its value from the mutant vector. $\boldsymbol{u}_i = [u_{i1}, u_{i2}, \ldots, u_{iD}]^T$ forms the trial vector.

7.3 Selection

The selection in original DE is greedy: the trial vector ($\boldsymbol{u}_i$) is compared with the target solution ($\boldsymbol{x}_i$), and the better one is selected as a population member of the next iteration.

7.4 Recent Variants

There has been many studies on analysis or development of new variants of DE by new initialization methods [3], (self-) adaption of the control parameters [10, 23], and reformulation or ensemble of the evolutionary operators (mutation, recombination, and selection) [33, 41, 43]. Several variants of DE for permutation problems have also been developed (e.g., for job-shop scheduling [71]). Many researchers have employed DE to optimize diverse practical problems raised in different fields of engineering. The readers are referred to a comprehensive review of DE [13, 14, 52] for detailed information about progress in this stream of EAs.

8 Other Relevant Methods

There are several other metaheuristic optimization methods that share the flexibility and stochastic nature of EAs. The most well-known group is swarm intelligence, which is an umbrella term for a variety of methods that mimic the behavior of a swarm of a species such as ants, fish, or bees. Some of the most well-known methods in this group are:

1. *Particle Swarm Optimization* (PSO): Inspired by a population of moving individuals such as flocks of birds, PSO was introduced by Kennedy and Eberhart [38]. In PSO, each particle represents an individual solution in D-dimensional search space. Particles are in motion, and their velocity is updated according to their current velocity (inertia), their best position so far (individual memory), and the current swarm best individual (swarm knowledge). This update process involves some randomness to improve exploration. PSO has gained a lot of interest among researchers and many variants of this method have been developed in the last 25 years (see [60] for a review).
2. *Ant Colony Optimization* (ACO): ACO [20, 21] is a population-based optimization method that has been developed mainly to find the optimal path through graphs, which makes is a desirable option for combinatorial problems such as the traveling salesman and vehicle routing problems. ACO was inspired by the strategy of ants foraging for finding the shortest path to food. The ants leave a pheromone trail on the ground which provides a means of communication. A stronger pheromone (associated with a better solution) attracts other ants more strongly. Several variants of ACO has been developed and applied to combinatorial optimization as well as problems in different disciplines of engineering and computer science [46].

There are other stochastic optimization methods that do not fit in the definition of swarm intelligence or EAs. For example, Simulated Annealing (SA) [39] was inspired by the annealing process in metallurgy. This single-point search method may accept a new solution that is worse than an existing one to improve global search. The probability of acceptance of a worse solution depends on the fitness difference and the temperature, which gradually reduces so that worse solutions are less likely to be accepted. Following this concept, several variants of SA for single and multi-objectivemulti-objective optimization optimization have been developed [65].

It is remarkable that there are many other metaheuristic methods which have been introduced more recently; however, the novelty of these method is usually limited to *metaphors*, not mathematical concepts [62]. The fast growth in the quantity of such methods has been criticized [24, 62], since these seemingly novel methods introduce new terminology (instead of mathematically novel concepts or formulations for their search operators).

9 Memetic Algorithms

Memetic algorithms (MAs) are a more recent extension of EAs. MAs utilize some form of local search as a mechanism for individual improvement. This mechanism can be provided by another heuristic or metaheuristic method. For example, the earliest memetic algorithm combined GA with SA for the traveling salesman problem [50]. MAs gain increased popularity based on implications of the *No Free Lunch* (NFL) theorem [69, 70], which states that any two algorithms are equivalently good if their performances are averaged over all possible problems. Since an MA can combine an EA with a heuristic and iterative local search, MAs are sometimes called hybrid evolutionary algorithms [47].

In MAs, the local search mechanism may activate after mutation and/or after recombination but before mutation. Since mutation is a disruptive operative, applying local search after mutation is a more rational choice.

The employed local search is generally a fast iterative (meta-) heuristic method, such as local tabu search [68] or a micro genetic algorithm (a GA with small population size and short evolution interval) [37]. It should be realised that the benefits of local search comes at the cost of extra function evaluations which could have been used for evolutionary operators. Therefore, an MA should decide on [48]:

1. the frequency/probability of performing local search.
2. the solutions to which local search should be applied.
3. the termination criteria or the maximum budget for local search (intensity of local search).

A more particular form of MA may employ a heuristic based on the existing knowledge in a specific application. For example, the hybrid method developed in [1] employs a heuristic search which was formulated solely based on engineering knowledge about the mechanical behavior of structures. This customized heuristic could make substantial changes to the sampled solution from the EA.

10 Summary and Conclusions

This chapter has provided an introduction for different streams of evolutionary algorithms (EAs) for parameter optimization. The advantages and disadvantages of EAs, when compared with classical optimization methods, were discussed. An overview of the basic concepts and formulations for four streams of EAs for parameter optimization have been provided. These streams are genetic algorithms (GAs), evolution strategies (ESs), evolutionary programming (EP), and differential evolution (DE). Finally, memetic algorithms, an extension to EAs, were reviewed.

Acknowledgements The research is supported by an Australian Research Council Discovery Project (DP190102637)

References

1. Ahrari, A., Deb, K.: An improved fully stressed design evolution strategy for layout optimization of truss structures. Comput. Struct. **164**, 127–144 (2016)
2. Ahrari, A., Kramer, O.: Finite life span for improving the selection scheme in evolution strategies. Soft. Comput. **21**(2), 501–513 (2017)
3. Ali, M., Pant, M., Abraham, A.: Unconventional initialization methods for differential evolution. Appl. Math. Comput. **219**(9), 4474–4494 (2013)
4. Bäck, T., Foussette, C., Krause, P.: Contemporary Evolution Strategies. Springer, Berlin (2013)
5. Back, T., Hammel, U., Schwefel, H.P.: Evolutionary computation: comments on the history and current state. IEEE Trans. Evol. Comput. **1**(1), 3–17 (1997)
6. Bäck, T., Rudolph, G., Schwefel, H.P.: Evolutionary programming and evolution strategies: similarities and differences. In: In Proceedings of the Second Annual Conference on Evolutionary Programming. Citeseer (1993)
7. Bartz-Beielstein, T., Preuss, M., Schmitt, K., Schwefel, H.P.: Challenges for Contemporary Evolutionary Algorithms. Citeseer, Princeton (2010)
8. Beyer, H.G., Schwefel, H.P.: Evolution strategies-a comprehensive introduction. Nat. Comput. **1**(1), 3–52 (2002)
9. Bianchi, L., Dorigo, M., Gambardella, L.M., Gutjahr, W.J.: A survey on metaheuristics for stochastic combinatorial optimization. Nat. Comput. **8**(2), 239–287 (2009)
10. Brest, J., Bošković, B., Greiner, S., Žumer, V., Maučec, M.S.: Performance comparison of self-adaptive and adaptive differential evolution algorithms. Soft. Comput. **11**(7), 617–629 (2007)
11. Chang, C.K., Christensen, M.J., Zhang, T.: Genetic algorithms for project management. Ann. Softw. Eng. **11**(1), 107–139 (2001)
12. Das, S., Maity, S., Qu, B.Y., Suganthan, P.N.: Real-parameter evolutionary multimodal optimization-a survey of the state-of-the-art. Swarm Evol. Comput. **1**(2), 71–88 (2011)
13. Das, S., Mullick, S.S., Suganthan, P.N.: Recent advances in differential evolution-an updated survey. Swarm Evol. Comput. **27**, 1–30 (2016)
14. Das, S., Suganthan, P.N.: Differential evolution: a survey of the state-of-the-art. IEEE Trans. Evol. Comput. **15**(1), 4–31 (2010)
15. Deb, K.: Introduction to evolutionary multiobjective optimization. Multiobjective Optimization, pp. 59–96. Springer, Berlin (2008)
16. Deb, K., Agrawal, R.B., et al.: Simulated binary crossover for continuous search space. Complex Syst. **9**(2), 115–148 (1995)
17. Deb, K., Joshi, D., Anand, A.: Real-coded evolutionary algorithms with parent-centric recombination. In: Proceedings of the 2002 Congress on Evolutionary Computation. CEC'02 (Cat. No. 02TH8600), vol. 1, pp. 61–66. IEEE (2002)
18. Deb, K., Myburgh, C.: Breaking the billion-variable barrier in real-world optimization using a customized evolutionary algorithm. In: Proceedings of the Genetic and Evolutionary Computation Conference 2016, pp. 653–660 (2016)

19. Deb, K., Myburgh, C.: A population-based fast algorithm for a billion-dimensional resource allocation problem with integer variables. Eur. J. Oper. Res. **261**(2), 460–474 (2017)
20. Dorigo, M.: Optimization, learning and natural algorithms. Ph. D. Thesis, Politecnico di Milano (1992)
21. Dorigo, M., Di Caro, G.: Ant colony optimization: a new meta-heuristic. In: Proceedings of the 1999 Congress on Evolutionary Computation-CEC99 (Cat. No. 99TH8406), vol. 2, pp. 1470–1477. IEEE (1999)
22. Eiben, A., Smith, J.: Popular evolutionary algorithm variants. Introduction to Evolutionary Computing, pp. 99–116. Springer, Berlin (2015)
23. Elsayed, S.M., Sarker, R.A., Essam, D.L.: An improved self-adaptive differential evolution algorithm for optimization problems. IEEE Trans. Industr. Inf. **9**(1), 89–99 (2012)
24. Fister Jr, I., Mlakar, U., Brest, J., Fister, I.: A new population-based nature-inspired algorithm every month: is the current era coming to the end. In: Proceedings of the 3rd Student Computer Science Research Conference, pp. 33–37. University of Primorska Press, Koper (2016)
25. Fogel, L.J.: Autonomous automata. Ind. Res. **4**, 14–19 (1962)
26. Fogel, L.J.: On the organization of intellect. Ph.D. thesis, University of California, Los Angeles–Engineering (1964)
27. Freitas, A.A.: A review of evolutionary algorithms for data mining. Data Mining and Knowledge Discovery Handbook, pp. 371–400 (2009)
28. Gao, K., Cao, Z., Zhang, L., Chen, Z., Han, Y., Pan, Q.: A review on swarm intelligence and evolutionary algorithms for solving flexible job shop scheduling problems. IEEE/CAA J. Autom. Sin. **6**(4), 904–916 (2019)
29. Gutin, G., Punnen, A.P.: The Traveling Salesman Problem and Its Variations, vol. 12. Springer Science & Business Media (2006)
30. Hansen, N., Finck, S., Ros, R., Auger, A.: Real-parameter black-box optimization benchmarking 2009: noiseless functions definitions. Technical report (2009)
31. Hansen, N., Ostermeier, A.: Completely derandomized self-adaptation in evolution strategies. Evol. Comput. **9**(2), 159–195 (2001)
32. Holland, J.H.: Outline for a logical theory of adaptive systems. J. ACM (JACM) **9**(3), 297–314 (1962)
33. Islam, S.M., Das, S., Ghosh, S., Roy, S., Suganthan, P.N.: An adaptive differential evolution algorithm with novel mutation and crossover strategies for global numerical optimization. IEEE Trans. Syst., Man, Cybern., Part B (Cybern.) **42**(2), 482–500 (2011)
34. Jin, Y.: Surrogate-assisted evolutionary computation: recent advances and future challenges. Swarm Evol. Comput. **1**(2), 61–70 (2011)
35. Joseph, V.R.: Space-filling designs for computer experiments: a review. Qual. Eng. **28**(1), 28–35 (2016)
36. Katoch, S., Chauhan, S.S., Kumar, V.: A review on genetic algorithm: past, present, and future. Multimedia Tools and Applications, pp. 1–36 (2020)
37. Kazarlis, S.A., Papadakis, S.E., Theocharis, J., Petridis, V.: Microgenetic algorithms as generalized hill-climbing operators for ga optimization. IEEE Trans. Evol. Comput. **5**(3), 204–217 (2001)
38. Kennedy, J., Eberhart, R.: Particle swarm optimization. In: Proceedings of ICNN'95-International Conference on Neural Networks, vol. 4, pp. 1942–1948. IEEE (1995)
39. Kirkpatrick, S., Gelatt, C.D., Vecchi, M.P.: Optimization by simulated annealing. Science **220**(4598), 671–680 (1983)
40. Koza, J.R., Koza, J.R.: Genetic Programming: On the Programming of Computers by Means of Natural Selection, vol. 1. MIT Press, Cambridge (1992)
41. Leon, M., Xiong, N.: Investigation of mutation strategies in differential evolution for solving global optimization problems. In: International Conference on Artificial Intelligence and Soft Computing, pp. 372–383. Springer, Berlin (2014)
42. Lian, Y., Oyama, A., Liou, M.S.: Progress in design optimization using evolutionary algorithms for aerodynamic problems. Prog. Aerosp. Sci. **46**(5–6), 199–223 (2010)

43. Mallipeddi, R., Suganthan, P.N., Pan, Q.K., Tasgetiren, M.F.: Differential evolution algorithm with ensemble of parameters and mutation strategies. Appl. Soft Comput. **11**(2), 1679–1696 (2011)
44. Mesghouni, K., Hammadi, S., Borne, P.: Evolutionary algorithms for job-shop scheduling. Int. J. Appl. Math. Comput. Sci. **14**(1), 91–104 (2004)
45. Metaxiotis, K., Liagkouras, K.: Multiobjective evolutionary algorithms for portfolio management: a comprehensive literature review. Expert Syst. Appl. **39**(14), 11685–11698 (2012)
46. Mohan, B.C., Baskaran, R.: A survey: ant colony optimization based recent research and implementation on several engineering domain. Expert Syst. Appl. **39**(4), 4618–4627 (2012)
47. Moscato, P., Cotta, C.: A gentle introduction to memetic algorithms. Handbook of Metaheuristics, pp. 105–144. Springer, Berlin (2003)
48. Neri, F., Cotta, C.: Memetic algorithms and memetic computing optimization: a literature review. Swarm Evol. Comput. **2**, 1–14 (2012)
49. Nguyen, T.T., Yang, S., Branke, J.: Evolutionary dynamic optimization: a survey of the state of the art. Swarm Evol. Comput. **6**, 1–24 (2012)
50. Norman, M.G., Moscato, P.: A competitive and cooperative approach to complex combinatorial search. In: Proceedings of the 20th Informatics and Operations Research Meeting, pp. 3–15. Citeseer (1991)
51. Ostermeier, A., Gawelczyk, A., Hansen, N.: Step-size adaptation based on non-local use of selection information. In: International Conference on Parallel Problem Solving from Nature, pp. 189–198. Springer (1994)
52. Pant, M., Zaheer, H., Garcia-Hernandez, L., Abraham, A., et al.: Differential evolution: a review of more than two decades of research. Eng. Appl. Artif. Intell. **90**, 103479 (2020)
53. Pelamatti, J., Brevault, L., Balesdent, M., Talbi, E.G., Guerin, Y.: How to deal with mixed-variable optimization problems: an overview of algorithms and formulations. World Congress of Structural and Multidisciplinary Optimisation, pp. 64–82. Springer, Berlin (2017)
54. Potvin, J.Y.: State-of-the art review – evolutionary algorithms for vehicle routing. INFORMS J. Comput. **21**(4), 518–548 (2009)
55. Rechenberg, I.: Cybernetic solution path of an experimental problem. Royal Aircraft Establishment Library Translation 1122 (1965)
56. Rechenberg, I.: Evolutionsstrategien. Simulationsmethoden in der Medizin und Biologie, pp. 83–114. Springer, Berlin (1978)
57. Salomon, R.: Re-evaluating genetic algorithm performance under coordinate rotation of benchmark functions. a survey of some theoretical and practical aspects of genetic algorithms. BioSystems **39**(3), 263–278 (1996)
58. Schwefel, H.P.: Kybernetische evolution als strategie der experimentellen forschung in der stromungstechnik. Diploma thesis, Technical University of Berlin (1965)
59. Schwefel, H.P.: Numerical Optimization of Computer Models. Wiley, Chichester (1981)
60. Sengupta, S., Basak, S., Peters, R.A.: Particle swarm optimization: a survey of historical and recent developments with hybridization perspectives. Mach. Learn. Knowl. Extr. **1**(1), 157–191 (2019)
61. Sigaud, O., Wilson, S.W.: Learning classifier systems: a survey. Soft. Comput. **11**(11), 1065–1078 (2007)
62. Sörensen, K.: Metaheuristics – the metaphor exposed. Int. Trans. Oper. Res. **22**(1), 3–18 (2015)
63. Storn, R., Price, K.: Differential evolution-a simple and efficient heuristic for global optimization over continuous spaces. J. Global Optim. **11**(4), 341–359 (1997)
64. Suganthan, P.N., Hansen, N., Liang, J.J., Deb, K., Chen, Y.P., Auger, A., Tiwari, S.: Problem definitions and evaluation criteria for the cec 2005 special session on real-parameter optimization. KanGAL Rep. **2005005**(2005), 2005 (2005)
65. Suman, B., Kumar, P.: A survey of simulated annealing as a tool for single and multiobjective optimization. J. Oper. Res. Soc. **57**(10), 1143–1160 (2006)
66. Tanabe, R., Ishibuchi, H.: A review of evolutionary multimodal multiobjective optimization. IEEE Trans. Evol. Comput. **24**(1), 193–200 (2019)

67. Valdebenito, M.A., Schuëller, G.I.: A survey on approaches for reliability-based optimization. Struct. Multidiscip. Optim. **42**(5), 645–663 (2010)
68. Vazquez, M., Whitley, L.D.: A hybrid genetic algorithm for the quadratic assignment problem. In: Proceedings of the 2nd Annual Conference on Genetic and Evolutionary Computation, pp. 135–142 (2000)
69. Wolpert, D.H., Macready, W.G.: No free lunch theorems for optimization. IEEE Trans. Evol. Comput. **1**(1), 67–82 (1997)
70. Wolpert, D.H., Macready, W.G.: Coevolutionary free lunches. IEEE Trans. Evol. Comput. **9**(6), 721–735 (2005)
71. Yuan, Y., Xu, H.: Flexible job shop scheduling using hybrid differential evolution algorithms. Comput. Ind. Eng. **65**(2), 246–260 (2013)

An Overall Characterization of the Project Portfolio Optimization Problem and an Approach Based on Evolutionary Algorithms to Address It

Eduardo Fernández, Efrain Solares, Carlos A. Coello Coello, and Victor De-León-Gómez

Abstract This chapter describes the main features of project portfolio selection and formalizes a problem statement that considers these features. We provide a simple but comprehensive illustrative example that shows the usefulness of the problem statement and argue that there are no published approaches so far that deal with its whole complexity. We also provide some guidelines to build such an approach based on the use of evolutionary algorithms.

1 Introduction

Resource allocation problems are ubiquitous in business and government organizations [53] because there usually are more good project proposals than resources available to support all of them. Of course, every organization must carefully select how to distribute its resources and which projects should they support to create the best impacts on the organization's objectives. Each of these project proposals can be defined as a collection of related and interdependent activities that have common specific goals. Business organizations often engage in resource allocation activities to increase profits or improve productivity (e.g., [10, 47, 54]) while government organizations do it trying to increase the quality of life of the society (e.g., [60]). Given a set of project proposals, a project portfolio (simply portfolio hereafter) defines if

E. Fernández · E. Solares (✉) · V. De-León-Gómez
Universidad Autónoma de Coahuila - FCA, Blvrd. Revolución, #151 Ote. 27000, Torreón, Mexico
e-mail: efrain.solares@uadec.edu.mx

E. Fernández
e-mail: eddyf171051@gmail.com

V. De-León-Gómez
e-mail: v.gomez@uadec.edu.mx

C. A. Coello Coello
Departmento de Computación del CINVESTAV-IPN, Av Instituto Politécnico Nacional #2508, 07360 Ciudad de México, Mexico
e-mail: ccoello@cs.cinvestav.mx

K. R. Harrison et al., *Evolutionary and Memetic Computing for Project Portfolio Selection and Scheduling*, Adaptation, Learning, and Optimization 26,
https://doi.org/10.1007/978-3-030-88315-7_4

each proposal is supported, or if it is not. The supported projects in the portfolio share resources during a given period, among which there may be complementarity, incompatibility, or synergies produced by sharing costs and benefits derived from conducting more than one project at the same time [25]. Project portfolio selection is the activity where a subset of candidate projects that (roughly) optimizes the organization's objectives is chosen. Each of these projects has a related cost and estimated impacts on the multiple objectives of the organizations. Thus, the selection procedure is performed under a constrained multi-criteria approach where the impacts of each project are contemplated to define the overall impacts of the portfolio, in such a way that the required costs of the supported projects do not exceed the available resources.

Project portfolio selection problems share some of the following characteristics: (i) considering multiple (sometimes many) and conflicting criteria [5, 37, 39]; (ii) modeling the *decision maker* (DM) preferences to select a convenient trade-off among the impacts on the criteria (e.g., [22, 56]); (iii) dealing with a set of constraints related to the available resources, where the budgetary constraint is only one of them (cf. [41, 46]); (iv) managing uncertainty (e.g., [9, 32, 43, 44, 57]); (v) finding the best portfolio through an optimization procedure (e.g., [22, 27, 31, 38]); (vi) handling synergy among projects (e.g., [54]). Usually, a combination of Multicriteria Decision Aiding (MCDA) and optimization methods is used. MCDA methods are used to aggregate the impacts on the criteria by considering the decision maker preferences, and the optimization methods are used to define the portfolio that maximizes the impact(s) on the decision maker objective(s).

Project portfolio selection has been addressed from many fields [39] such as selection of infrastructure projects [33, 34, 56], selection of renewable energies [26, 29, 50], selection of products to include in a manufacturing process [19, 59], and selection of Research and Development (R&D) projects [3, 4, 35, 36]. Particularly, selection of R&D projects is a very active research area.

As detailed in Sect. 2, most of the published works that address project portfolio selection are subject to at least one of the following issues: (i) poor modeling of decision maker preferences; (ii) use of simplistic approaches that neglect relevant situations; (iii) neglecting interaction between projects; (iv) selecting portfolios with projects performing very well on some criteria but drastically underperforming on some other criteria; (v) lack of capacity to create portfolios that are consistent with the ranking of individual projects or a set of ordered classes; (vi) not defining a sufficiently small number of portfolios as the best alternatives, (vii) poor modeling of uncertainty and realistic considerations. Therefore, the main goal of this work is to describe an overall approach that effectively deals with all these issues.

The chapter is structured as follows. Section 2 describes how the procedure of portfolio selection has been historically carried out by different approaches and provides a discussion about their advantages and limitations. Section 3 formally describes the project portfolio selection problem and provides an illustrative example. Section 4 describes our proposals to improve traditional methods and explains the most plausible features that a novel approach which incorporates such proposals should fulfill. Finally, Sect. 5 provides our conclusions and some possible paths for future research.

2 A Review of the Project Portfolio Selection Process

A review of historical and state-of-the-art approaches to the project portfolio selection process is described here. Section 2.1 mentions how some of these approaches are used in the main stages of such a process. Section 2.2 characterizes the approaches and provides arguments about their advantages and limitations, as well as when they should be used.

2.1 Phases in the Project Portfolio Selection Process

Generally, the project portfolio selection process consists of two phases: the evaluation of individual projects and the creation of the portfolio.

In the first phase, the criterion performances of each project are evaluated, and its resource requirements are specified. Frequently, a project score is calculated using a certain multicriteria aggregation method (e.g., [1, 40]). The most popular one consists in a weighted sum function $\sum w_i f_i$, where w_i reflects the importance of the ith criterion, and f_i is the value that the DM assigns to the performance of the ith criterion (cf. [1, 49]). The main advantage of the weighted sum is its simplicity and its properties of transitivity, comparability, and independence with respect to irrelevant alternatives, which are convenient for creating a ranking of projects. Its main drawbacks are that it admits total compensation (it cannot consider veto effects), it requires a constant trade-off rate (for each pair of criteria, a magnitude for which one of the criterion degrades is compensated for a fixed magnitude for which the other improves), it assumes no interaction among criteria, it demands cardinal information in f_i, and it does not admit threshold effects. The score can be assigned using other multicriteria aggregation methods such as Preference Ranking Organization METHod for Enrichment of Evaluations (PROMETHEE; e.g., [43, 58]), capable of handling ordinal and qualitative information, as well as threshold effects, without requiring the constant trade-off rate; however, the ranking derived from PROMETHEE lacks independence with respect to irrelevant alternatives.

This first phase usually ends when at least one of the following goals is met: (i) a multicriteria description of each project is available, so that its potential contribution to the portfolio can be added to that of other projects; (ii) there is a "value" associated with each project and a ranking of them; (iii) projects that do not meet minimum acceptability conditions are detected, and therefore are excluded from the second phase; (iv) projects are assigned to certain ordered classes (for example, "should be supported", "doubt", "should not be supported"). Goal (iv) is particularly hard to achieve because it requires expression of preferences from the DM on potentially many projects; even in the presence of few projects it can also be hard if the DM is an entity composed of multiple individuals or if the structure of criteria describing the projects is complex.

In the second phase, where a portfolio is built, it is decided which projects will be supported on the basis of the information obtained in the first phase and taking into account the restrictions that the DM considers convenient. Most of the methods used in this phase belong to one of the following general approaches:

1. Methods that use the results of a multicriteria evaluation of projects as the main input (e.g., [28, 30, 55]). The results sorts the projects in a ranking or in a set of categories; thus, the methods exploiting these results are of two types: (i) the portfolio is built respecting the order of the ranking of projects, until resources are exhausted; (ii) the requirements of the projects in the category "must be supported" are adjusted so as to reach the available resources [23]. These methods face difficulties when there are the so-called segmentation constraints (e.g., [41]). These are restrictions of the type "30% of the budget must be invested in basic science projects." In these cases, a heuristic is necessary to satisfy the constraints with minimal ranking violations. Some advantages of this type of methods are their simplicity (when there are no segmentation constraints) and that they guarantee that the "best" projects belong to the portfolio. However, no portfolio measure is optimized, the methods only get a feasible portfolio composed of projects that are individually good.
2. Methods that use a 0–1 programming to exploit the scores of the projects (e.g., [42]). Like a knapsack problem, these methods optimize the total "value" of the portfolio within the imposed restrictions; this value is calculated by adding the scores of the projects that it contains. Some advantages of these methods are: (i) an optimal portfolio is obtained from the point of view of the calculated "value"; (ii) high computational efficiency that allows addressing problems with a large number of projects (even tens of thousands); (iii) possibility of imposing segmentation constraints. Some disadvantages are: (a) generally, the optimal solution does not follow the ranking priorities; high-quality projects can be left out of the portfolio when there is a combination of lower-quality projects with lower resource consumption that contributes more to the "value" of the portfolio (e.g., [41, 58]); (b) as stated by Barbati et al. [6], this approach does not take into account portfolio balance requirements, with the risk of selecting a portfolio which is composed of projects with good evaluations on the same criterion or on the same small subset of criteria, and rather poor evaluations in the remaining criteria; (c) handling synergetic effects among subsets of projects is very difficult; (d) the "value" associated with each project must be cardinal.
3. If each project can be described by a vector $(v_1, \ldots, v_N)$ of "benefits" or positive impacts, and if there is a way to aggregate those project-level impacts into a vector $(V_1, \ldots, V_N)$ at portfolio-level, the best portfolio is then obtained by maximizing $(V_1, \ldots, V_N)$ within the region defined by the problem constraints, including the segmentation constraints (e.g., [8, 15]). Some advantages of this approach are:

(i) the vector $(V_1, \ldots, V_N)$ provides the level of achievement of the organization in its different objectives; the DM can naturally express his/her portfolio preferences and select the final solution according to his/her objectives, directly; the approaches in points 1 and 2 only allow this, at most, indirectly; (ii) balanced portfolios can be obtained; (iii) the way of adding the v_i values can include models of positive and negative synergy effects (e.g., [54]). Some disadvantages of the approach are: (a) as the number of objectives increases, the computational complexity increases, and metaheuristics are required to solve problems in more than a few dozen proposals; (b) the number of non-dominated portfolios grows exponentially with the dimensionality of the problem, which complicates the process of identifying the best solution; (c) the best projects are not guaranteed to form part of the portfolio, as there may be a combination of lower quality projects that contribute more to the vector $(V_1, \ldots, V_N)$.

In summary, most of the existing approaches for selecting projects and portfolios suffer of some of the following drawbacks:

- Rough modeling of the decision maker's preferences when multiple criteria are aggregated to obtain some "value" measure of the projects. This includes at least one of the following criticisms: (i) poor handling of ordinal and qualitative information; (ii) no consideration of interaction among criteria; (iii) full compensation and no consideration of threshold effects; (iv) difficulty to address imperfect information on criterion performance and decision model parameters. In case of R&D projects, (v) no consideration of hierarchical structures; (vi) no explicit consideration of the project's probability of success.
- The portfolio is selected by using a very simple heuristic; no portfolio is optimized, and segmentation constraints are very difficult to be fulfilled.
- No consideration of interacting projects when the portfolio is optimized.
- The optimization process leads to unbalanced portfolios, in which many projects may have good performance in the same criteria and a rather poor one in other subset of criteria.
- The optimization process does not guarantee that the best projects are in the final portfolio; that is, the order of quality of projects is violated by the optimization.
- Too many non-dominated characterized portfolios are obtained as Pareto solutions, and hence the DM cannot identify the best solution (or requires a great cognitive effort). This is even more relevant when the DM is a group with, to a certain extent, conflicting views.
- Imperfect information on project requirements and available resources are not appropriately modelled.

2.2 Characterizing a Plausible Project Portfolio Selection Approach

In addition to the typical project-level constraints (e.g., resource constraints), there are constraints on the conformity of the DM with the selected portfolio that must also be considered. For example, the DM can be reluctant to include in the portfolio projects with low performance in one or more criteria.

Another example is that, if the portfolios are composed by R&D projects, regarding the approach of Point 3 and $(v_1, \ldots, v_N)_k$ representing the vector of potential impacts by the kth project, to make a meaningful aggregation of $(v_1, \ldots, v_N)_k$ for all the projects belonging to the portfolio, each project should have a relatively high probability to reach its potential impacts. This condition can be imposed by discarding the subset of projects that do not reach an acceptable high level of probability of success. Alternatively, $(v_1, \ldots, v_N)_k$ may represent averaged values of the kth project impacts, calculated by using the setting of probability of success.

Other conformity constraints are: (i) demanding that the projects are assigned to a minimum acceptability class at the level of any non-elementary criterion within the hierarchy of criteria (for example, only projects with a "High" scientific impact); (ii) the portfolio must minimize the number of supported projects with quality inferior to that of others that are not supported; (iii) the portfolio must maximize the number of projects assigned to the best classes; (iv) segmentation constraints. These conformity constraints can be defined as "strict" or "soft" constraints, according to the preferences of the DM and the assessments can be performed at the level of any non-elementary criterion.

In large enterprises or public organizations, the entity in charge of the decision may be a group, in which, usually, its members disagree concerning the parameter values of the decision model; or the DM may be an inaccessible person (e.g., the CEO of a multinational enterprise), in which case her/his representative should set the model's parameters, and handle the related arbitrariness, imprecision, uncertainty, and ill-determination. These types of imperfect knowledge are usually also involved in the definition of the criteria impacts.

A plausible approach to deal with the complete process of project evaluation and portfolio selection should fulfill the following features:

- To use a hierarchical decomposition of the evaluation problem; the root objective is the quality of the project. In the case of R&D, quality depends on sub-criteria such as impact, probability of success, and consumption of resources; the impact can contain different dimensions, according to the objectives of the organization. The probability of success depends on several sub-criteria. The possibility of different interactions between criteria is not ruled out.
- The multi-criteria decision model must allow evaluating the general quality of the project, as well as any of the non-elementary criteria (for example, the probability of success in case of R&D, the impact on healthcare or educational services in case of social-oriented projects), so that, at the time of forming the portfolio, the presence of certain classes of projects may be prohibited, limited, or encouraged

(for example, "only good quality projects," or "only projects with a high probability of success"). These class assignments must reflect the preferences of the decision maker.
- The multi-criteria preference model should be able to handle: (i) ordinal or qualitative information; (ii) interaction between criteria; (iii) veto and no-comparability effects; (iv) inaccurate or uncertain information on criteria values; (v) imperfect information in the decision model parameters, due to conflicting opinions of a group of decision makers, or the inaccessibility of the top decision maker.
- Optimize the quality of the portfolio, considering: (i) its potential impacts; (ii) the feasibility of obtaining them; (iii) that it only contains projects of certain classes; (iv) that it does not contain projects of clearly inferior quality than others that are not supported; (v) positive and negative synergy effects; (vi) segmentation constraints; (vii) requirements from the DM regarding balanced portfolios.
- The optimization method must be able to handle imprecise and uncertain information on budget requirements, on the amount of available resources, and on portfolio-level criteria. Additionally, it must be able to solve problems with hundreds of applicant projects.
- The required cognitive effort of the top DM or his/her delegates to reach a near-optimal solution must be kept within tolerable limits.

To the best of our knowledge, there are no prior published works that consider all these features. This may be due to the high difficulty of dealing with an overall model that fulfills all of them. However, the modern complexity of real-life situations and the specific needs of some contexts demand greater efforts in achieving such a model. We describe here an overall project portfolio selection approach that considers all the features and can overcome all the issues described in Sect. 2.1. Particularly, the described approach provides flexibility along the whole decision-making process to incorporate the decision maker's preferences and aspirations about how the best solution should be built.

3 Problem Statement

The problem that is of interest for this work is described here. Section 3.1 is intended to provide insights into the problem. Section 3.2 presents a toy example and, finally, Sect. 3.3 formalizes the problem through the use of mathematical notation.

3.1 Problem Description

Typically, the senior management or its representative (maybe a single person or a group; hereafter the DM) of an institution is in charge for deciding how to allocate the institution's resources among the project proposals and select a final project portfolio.

Achieving this implies assessing individual projects on the one hand and potential portfolios on the other. Sometimes, the DM will prefer the selected portfolio to be made up of the highest-ranked or best evaluated projects and avoid discarding them to incorporate the lower-ranked or worst evaluated ones. However, most of the time the DM will be interested in a trade-off between achieving this general goal and increasing the portfolio performances. Thus, to define the extent to which the order of quality of projects should be respected during portfolio selection, the DM needs to determine what he/she considers to be the best trade-off between respecting the ranking or evaluation of projects and maximizing the performances of the selected portfolio.

At both levels of assessment, the DM is often concerned about the impacts of both projects and portfolios on a set of criteria that can be naturally structured as a hierarchy. A project portfolio selection approach must consider both ordinal and cardinal scales, deal with heterogeneity related to the nature of the impacts, handle constraints at project level and regarding the conformity of the DM with the portfolio, cope with imperfect knowledge, and allow pairs of criteria, as well as pairs of projects, to interact. The imperfect knowledge is originated by one or more of the following causes:

- Imprecision defining values of preferential parameters such as criteria weights or veto thresholds (e.g., when senior management is composed of multiple individuals);
- Ambiguity establishing the criteria impacts (e.g., when multiple experts express their estimations);
- Uncertainty on project costs;
- Uncertainty on actual resources required.

The aggregation of criteria impacts must be able to be exploited on several possible ways (note that the DM may be interested in more than one): (i) selecting the best projects/portfolios; (ii) ranking the projects/portfolios from the most preferred to the least preferred; (iii) assigning the projects/portfolios to one of several preferentially ordered classes (e.g., good, average, bad). Note that, even when the overall quality is the main concern, the DM may also be interested in the projects' or portfolios' evaluations/ranks regarding any other non-elementary criterion; thus, points (ii) and (iii) should be possible at any level of the hierarchy. For example, the DM can be interested in defining if a given project or portfolio has a "high" impact or level of feasibility.

Some goals that the DM would like to achieve by exploiting the aggregation procedure at a project level are:

- Discarding overall bad projects;
- Discarding projects that perform poorly in one or more other non-elementary criteria;
- Guaranteeing a minimal impact on one or more criteria.

Similarly, at a portfolio level, the goals are:

- Ensuring to a desired extent that the selected portfolio respects the ordering from the quality of projects;
- Neglecting portfolios that are not at least "acceptable" (e.g., from classes "bad", "acceptable", "good") or equivalent from an overall perspective or from a specific perspective;
- Avoiding to select portfolios with poor performance on some criteria.

The general problems that the approach described in Sect. 4 can deal with are characterized by one or more of the following situations, in which the DM can be interested in:

- Performing a trade-off between the impact of the portfolio on the overall criterion and ensuring that there are no globally better projects left out of the portfolio while globally worse projects are supported;
- Performing a trade-off between the impact of the portfolio on the overall criterion g_0 and ensuring that there is no project left out of the portfolio that was assigned to a better class than any other supported project.
- Discarding projects whose class is not acceptable in one or more non-elementary criteria;
- Discarding portfolios whose class is not acceptable in one or more non-elementary criteria;
- Considering only projects whose impacts on given criteria are at least a chosen value;
- Demanding that the built portfolio provides minimum impact levels on given criteria;
- Performing a portfolio optimization;
- Dealing with collective entities making decisions on projects and portfolios;
- Considering project interaction;
- Considering criteria interaction;
- Contemplating incomparability, veto conditions and preference thresholds;
- Dealing with imperfect knowledge whose origin is imprecision/uncertainty/ arbitrariness/vagueness on preferential parameters and criteria impacts as well as costs or required resources;
- Dealing with segmentation constraints.

Note how the DM would like to see the impacts of all the projects before defining what are the impact values for which he/she would neglect a project.

3.2 *An Illustrative Example*

Consider the case of a large governmental organization whose main interests are to increase the wellness of society. The organization has issued calls for proposals that positively impact on these interests. The organization's senior management (as the DM) decided that the supported projects should be those that better improve the country's Human Development Index (cf. [7, 48]). The Human Development Index (HDI) "measures the average achievements in a country in three basic dimensions of human development: longevity, knowledge and a decent standard of living," and provides a reasonable way to measure the impact on the society's wellness. However, the DM has realized significant biases between wellness of different social classes, so he/she wants to take special care of the lower social class.

The HDI is calculated using three factors: life expectancy index (LEI), educational attainment index (EAI), and real Gross Domestic Product per capita index (GDPI), such that HDI $=$ (LEI $+$ EAI $+$ GDPI)$/3$ [7]. The LEI factor is normalized at a maximum age of 85 years and a minimum of 25; thus, LEI $=$ (life expectancy $-25)/(85-25)$. The EAI factor is calculated from the Adult Literacy Index (ALI) that indicates the proportion of literate adults, and the combined primary, secondary and tertiary Enrollment Ratio Index (ERI) that indicates the proportion of children in these levels that really go to school; therefore, the factor is defined as EAI $=$ (2ALI $+$ ERI)$/2$. Note that it is normalized. To improve the LEI factor, the DM issues calls for health projects. The proposals can be: (i) medical care infrastructure (construction of hospitals, clinics, or equipment); (ii) health care programs (e.g., programs against chronic diseases, vaccines). Since here the goal is to reduce mortality, each project must estimate its impact on life savings that it can generate. For political reasons, the DM is more interested in the impact in the short term, although reducing mortality in the medium term should be also considered. The vocation of the DM for the lower social class requires differentiating saving lives between two different social classes. Another criterion at the portfolio level is the coverage of hospital care in the region. Finally, a conformity objective at portfolio level related to the LEI is to approximate a certain percentage of the budget on infrastructure projects.

Regarding the EAI, the DM issues calls for project proposals of two types: (i) adult literacy, and (ii) incorporation of children and young people into schools. Adult literacy projects can be infrastructure (schools, including staff) or services (online education, books). Each project must estimate two criteria, which are the numbers of adults who will become literate in the short and medium term (two criteria). The projects of incorporation of children and young people into schools can be of several types: (i) infrastructure (schools, including staff); (ii) services (free educational material, online education, double session); (iii) scholarships for low-income students. Each project must estimate two criteria, which are the numbers of students that it will contribute to incorporate or maintain into school in the short and medium term (two criteria). At the portfolio level, educational coverage in the region is another criterion.

Based on these criteria, each project can be assigned to one of the following quality assessment classes:

- High quality
- Average quality
- Poor quality

The projects assigned to the poor-quality class are neglected during the portfolio selection process. The criteria at the portfolio level are as follows:

Impact criteria:

- Number of lives estimated to be saved in the short and medium term, by social class;
- Number of adults who become literate in the short and medium term;
- School coverage that is achieved in the short and medium term;
- Number of projects that make strong synergy;
- Number of projects that make moderate synergy.

Conformity criteria:

- Minimize the number of projects with harmful environmental impact;
- Minimize the number of times that the quality assessment is violated in the portfolio, that is, the number of projects supported by the portfolio that are assigned to a worse quality class with respect to non-supported projects;
- In the health-related budget, approximate a minimum percentage provided to infrastructure;
- In the education-related budget, approximate a minimum percentage provided to infrastructure;
- From the general budget, try to reach or exceed a percentage that is dedicated to health;
- The DM wants to achieve balanced health and education rates, so another objective is to minimize |LEI – EAI|.

All the projects are grouped into four types according to their geographic region, and the DM is interested in providing at least 15% of the resources to each region.

The entity (as the DM) in charge of deciding how to allocate the organization's resources is made up of a set of individuals; the points of view of the individuals are rather conflicting so, it is not possible to define precise values for preference parameters. The DM is facing the problem of selecting a set of social projects among one thousand proposals, whose assessments have already been performed by experts. Multiple experts were requested to assess the projects and some of the experts provided vague estimations; therefore, imprecise requirements of resources were defined for each project, and a plurality of project scores have been defined for each criterion. This imprecision is also reflected at the portfolio-level criteria.

3.3 Problem Formalization

For clarity, we formalize the problem using a two-levels perspective. First we characterize projects and later we characterize portfolios. We use such characterizations to define the multi-objective optimization problem.

3.3.1 Project Level

Assume the existence of a set of project proposals, $P_r = \{p_1, p_2, \dots, p_N\}$, such that each $p_j \in P_r$ is assessed on a set of criteria $H = \{h_0, h_1, \dots\}$ structured as a hierarchy, where h_0 is the root-node denoting the global criterion. H_a is the set of all the criteria descending from h_a. E_h is the set of elementary criteria in the hierarchy. Defining and handling $h_i(p_j)$, $h_i \in H$, usually requires coping with imperfect knowledge.

Assume the existence of a set $B^a = \{B_1, B_2, \dots, B_{M_a}\}^a$ of classes, ordered with increasing preferences, to which a given project can be assigned regarding a non-elementary criterion h_a; that is, a project $p \in P_r$ can be assigned to a class $B_i \in B^a$ considering the performances of p on the criteria in H_a as well as the interaction between pairs of these criteria. The assignment of project p to class B_i regarding the criteria in H_a is denoted as $p \rightarrow_a B_i$, and the process of assigning projects to ordered classes is called sorting. Note that the difficulty of assigning projects to classes can lead to the necessity of an MCDA method to support the DM in this task.

Projects can be both ranked and sorted regarding the criteria in H_a for any non-elementary criterion h_a. It is of special relevance the ranking and sorting of projects from the perspective of h_0. Let $R^0 = \{p'_1, p'_2, \dots, p'_N\}$ be this ranking such that p'_i is not worse than p'_j if $i < j (p'_i, p'_j \in P_r)$, and $S^0 = \{B_1, B_2, \dots, B_{M_0}\}^0$ be this sorting such that p_i is not worse than p_j if $p_i \in B_k$ and $p_j \in B_l$ with $k \leq l$.

3.3.2 Portfolio Level

Let A be the set of portfolios, where each $x \in A$ is a vector $[x_1, \dots, x_i, \dots, x_N]^\top$ characterized by $x_i = \{0, 1\}$ denoting if the ith project is selected ($x_i = 1$) or not ($x_i = 0$). Each $x \in A$ is assessed on a pair of disjoint sets of impact criteria, $G = \{g_1, \dots, g_n\}$, and conformity criteria, $F = \{f_1, \dots, f_o\}$. Note that defining and handling $g_i(x)$ usually requires coping with imperfect knowledge, although it is usually not the case for $f_j(x)$. Also note that, generally, $G \cap E_h \neq \emptyset$ but $G \neq E_h$. A portfolio x can be assigned to one of a set of preferentially ordered classes $C = \{C_1, C_2, \dots, C_L\}$ regarding the criteria in G; $x \rightarrow C_i$ denotes that portfolio x has been assigned to class C_i.

Maximizing the performance of portfolio x on the impact criteria must not necessarily respect R^0 or S^0; meaning that, to increase $g_i(x)$, $g_i \in G$, the search procedure can set $x_j = 0$ and $x_l = 1$ even when $j < l$ $(p_j, p_l \in R^0)$ or $p_j \in B_k$ and $p_l \in B_p$ with $k < p$, what we call violation of the ranking or the sorting, respectively. Of course, this can be non-satisfactory for the DM and a minimization of the number of violations can be required. Let $Nv(x, R^0)$ and $Nv(x, S^0)$ be such numbers.

When the constraints related to the conformity of the DM with the selected portfolio are soft, these constraints can be introduced in the multi-objective problem formulation as objective functions. Let $Nv(x, SGMN)$ be the number of violations regarding the set of segmentation constraints, $SGMN$. Each element in the set $SGMN$ is a constraint whose definition is context-dependent; for example, for the constraint "at least 30% of the budget must be invested in basic science projects," $\sum y_i x_i \geq 0.30$ should be fulfilled, where $y_i = 1$ if the ith project is a basic science project and $y_i = 0$, otherwise. If it is not fulfilled, then it is counted as a violation and $Nv(x, SGMN)$ is increased by one. Similarly, if the DM defined that the portfolio must maximize the number of projects assigned to the best classes, then a measure such as $N_{best}(x) = \sum z_i x_i$, can be added to the problem formulation as an objective function, where $z_i = 1$ if $x_i \rightarrow_a B_{M_a}$ and $z_i = 0$, otherwise for any desired $h_a \in H/Em$. Therefore, $Nv(x, R^0)$, $Nv(x, S^0)$, $Nv(x, SGMN)$, $N_{best}(x) \in F$; all with decreasing direction but the last one.

3.3.3 Multi-objective Formulation

Let $U = \{(h_a, m_a)$: the DM wants to neglect any $p \in P_r$ when $p \rightarrow_a B_m$ and $m < m_a(h_a \in H/E_h, m_a \in \{2, 3, \ldots, L_a\}, B_m \in B^a)\}$ and $V = \{(g_i, l_a)$: such that the DM wants to neglect any $x \in A$ when $x \rightarrow C_l$ and $l < l_a(g_i \in G, l_a \in \mathbb{N}, C_l \in C)\}$ be the sets establishing if for a given non-elementary criterion h_a (respectively, g_i), the DM defines a category B_{m_a} (resp., C_l) as the minimum acceptability level for each project (resp., portfolio). Similarly, and without loss of generality, let $W = \{(h_i, t_i)$: the DM wants to neglect any $p \in P_r$ when $h_i(p) < t_i$ $(h_i \in H, t_i \in \mathbb{R})\}$ and $Z = \{(g_i, s_i)$: the DM wants to neglect any $x \in A$ when $g_i(x) < s_i$ $(g_i \in G, s_i \in \mathbb{R})\}$ be the sets establishing if for a given non-elementary criterion h_i (respectively, g_i), the DM defines a value t_i (resp., s_i) as the minimum acceptability impact for each project (resp., portfolio). Finally, each project $p_j \in P_r$ has an associated cost $c_j \in \mathbb{R}^+$ and the available budget is denoted by T.

Let $g_i(x) \in G$, and $f_j(x) \in F$; $i = 1, \ldots, n$; $j = 1, \ldots, o$. An example of the general statement of the project portfolio optimization problem is denoted, without loss of generality, by:

$$\text{Maximize}(g_1(x), \ldots, g_n(x), f_1(x), \ldots, f_o(x)) \tag{1}$$

Subject to

$$\begin{aligned}
&\sum c_i x_i \leq T \\
&\forall (h_a, m_a) \in U : p_i \rightarrow_a B_m \wedge m < m_a \Rightarrow x_i = 0; \\
&\neg\exists (g_a, l_a) \in V : x \rightarrow_a C_l \wedge l < l_a; \\
&\forall (h_i, t_i) \in W : h_i(x_i) < t_i \Rightarrow x_i = 0; \\
&\neg\exists (g_i, s_i) \in Z : g_i(x) < s_i; \\
&x_i = \{0, 1\}; \\
&i = 1, 2, \ldots, N.
\end{aligned}$$

This formulation must be adapted to the DM's requirements. For example, it makes sense to consider only one of the minimization objectives $Nv(x, R^0)$ and $Nv(x, S^0)$.

4 An Overall Approach to Project Portfolio Selection

The types of constraints and number of objective functions make it very difficult for exact optimization techniques to address Problem (1). The necessity to cope with imperfect knowledge on $g_i \in G$ increases the problem's complexity. Therefore, the characteristics of multi-objective evolutionary algorithms (MOEAs) are very convenient to address Problem (1). From such characteristics, it outstands their ability to obtain an approximation of the Pareto frontier in a single algorithm's run, their robustness regarding the shape of the Pareto front, and their ability to adapt to the specific requirements o f the DM.

The proposal of this work to build an overall approach to project portfolio selection consists of several stages that are described throughout this section.

4.1 Framework of the Approach

The inputs to the approach consist of the project's performance, the DM's preferential information, and the portfolio constraints and objectives, see Fig. 1. The project's performance consists, for example, of: (i) criteria scores of the projects; (ii) ranking(s) of the projects regarding specific non-elementary criteria; (iii) sorting(s) of the projects regarding specific non-elementary criteria. The approach uses the DM's preferences (e.g., criteria weights and preference thresholds, which can be imprecisely defined) to aggregate the project performances through a multi-criteria decision aiding method, and an evolutionary algorithm to optimize the portfolio-level objective functions fulfilling the portfolio-level constraints (such as the conformity objectives and conformity constraints). The preferential information allows the approach to create balanced portfolios considering, for example, the trade-off between high impacts on the portfolio-level objective functions, and keeping, as much as possible, compatibility with a ranking or sorting of projects.

In the following subsections, we provide insights on plausible components of the approach.

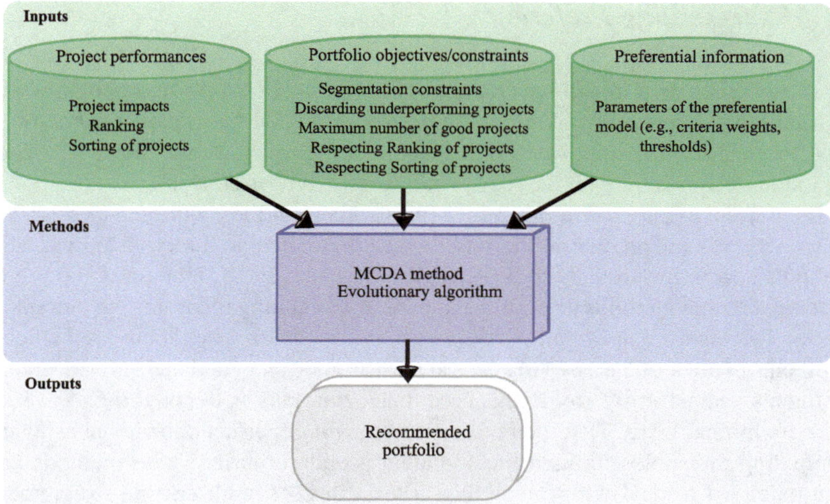

Fig. 1 Framework of an overall approach

4.2 *Coping with Imperfect Information on the Criteria Impacts*

Imperfect information (i.e., imprecision, arbitrariness, uncertainty and/or ill-determination) on the definition of elementary criteria can be effortlessly modeled by interval-numbers (e.g., [17]) and/or by pseudo-criteria that use the concept of discriminating thresholds (e.g., [52]) depending on the criteria's domain.

An interval number, $\boldsymbol{I} = [I^-, I^+]$, represents an imperfectly known real quantity r that is assumed to be between a lower limit I^- and an upper limit I^+ [45]. Therefore, any real number is a special case of an interval number for which $I^- = I^+$, and any $r \in [I^-, I^+]$ is called *realization* of $\boldsymbol{I}$. This paper uses boldface symbols to represent interval numbers. There is some basic arithmetic that can be used with interval numbers such as addition and subtraction; if $\boldsymbol{I} = [I^-, I^+]$ and $\boldsymbol{J} = [J^-, J^+]$, then $\boldsymbol{I} + \boldsymbol{J} = [I^- + J^-, I^+ + J^+]$ and $\boldsymbol{I} - \boldsymbol{J} = [I^- - J^+, I^+ - J^-]$. Furthermore, an order can be established between them $(I > J)$ with the following function:

$$P(\boldsymbol{I} \geq \boldsymbol{J}) = \begin{cases} 1 & \text{if } p_{IJ} > 1 \\ p_{IJ} & \text{if } 0 \leq p_{IJ} \leq 1 \\ 0 & \text{if } p_{IJ} < 0 \end{cases} \tag{2}$$

where $p_{IJ} = \frac{I^+ - J^-}{(I^+ - I^-) + (J^+ - J^-)}$, and $(I^+ = I^- = i \wedge J^+ = J^- = j) \Rightarrow P(\boldsymbol{I} \geq \boldsymbol{J}) = \begin{cases} 1 & \text{if } i > j \\ 0 & \text{otherwise} \end{cases}$

4.3 Representing Preferences

In practical scenarios, the dimensionality of Problem (1) is very high, complicating the search procedure and making it impractical for the DM to issue a final decision. Therefore, an *a priori* multi-criteria decision aiding method that allows the search technique to concentrate on the region of interest (ROI) is convenient.

The outranking approach inherently fulfills many of the requirements described in past sections and an outranking-based method can help to direct the optimization method's search towards the ROI. PROMETHEE and ELECTRE (cf. [51]) are the most outstanding methods that build and exploit outranking relations. An outranking relation between actions (e.g., portfolios) x and y occurs when there are sufficient arguments in favor of the assertion "x is at least as good as y" and there are not enough arguments against it; we say in this case that x outranks y, denoted by xSy. Some of the traditional ELECTRE methods can cope with imperfect knowledge using discriminating thresholds through the so-called pseudo-criteria. These methods have been extended to deal with interacting criteria in [24], with criteria structured as a hierarchy in [12] and with hierarchical assessments of action performances on interacting criteria in [11]. Corrente et al. [13] presented a multiple criteria hierarchy process to assign actions to preferentially ordered classes (i.e., sorting actions). Fernández et al. [17] proposed an interval-based outranking approach and a sorting method that can cope with imperfect knowledge in the form of interval numbers; this imperfect knowledge can be originated from preferential parameters (e.g., imprecise criteria weights) or from ill-determined, imprecise, uncertain, arbitrary (even missing) criterion impacts. Two improved methods based on the interval outranking for multi-criteria ordinal classification were presented by Fernández et al. [18].

The interval-based outranking method does not consider a multi-level hierarchy but the traditional structure of N criteria in a single hierarchical level; it uses an interval number to assess the performance of portfolio x on the ith perspective, creating a criterion that is defined as $\boldsymbol{g}_i(x) = [g_i^-(x), g_i^+(x)]$; where $g_i^-(x)$ and $g_i^+(x)$ are the minimum and maximum attainable values of the portfolio's performance on the corresponding perspective. These values should be defined according to the context; for example, as the minimum and maximum values of the scale of criterion g_i. The preferential parameters used by this method per criterion g_i are: (i) a criterion weight $\boldsymbol{w} = [w_i^-, w_i^+]$, denoting the importance assigned by the DM to g_i; (ii) a veto $\boldsymbol{v} = [v_i^-, v_i^+]$, representing the minimum difference between the performance of y over x on g_i that vetoes xSy; (iii) a threshold $\boldsymbol{\lambda} = [\lambda^-, \lambda^+]$ reflecting a threshold for a sufficient strength of criteria supporting xSy. Then, a partial outranking S_i between portfolios x and y is associated to each criterion g_i such that xS_iy occurs when there is high credibility that x outranks y regarding g_i. The credibility index $\sigma_i(x, y)$ of the assertion "x is at least as good as y regarding criterion g_i" is given by $\sigma_i(x, y) = P(\boldsymbol{g}_i(x) \geq \boldsymbol{g}_i(y))$, where $P(\cdot)$ is defined by Eq. (2). The δ-concordance coalition, $C(xS_\delta y)$, is the set of criteria g_i such that $\sigma_i(x, y) \geq \delta$ for a given credibility threshold δ. All criteria that do not belong to $C(xS_\delta y)$ are then a component of $D(xS_\delta y)$, the δ-discordance coalition. The concordance index, denoted by $\boldsymbol{c}(x, y, \delta) = [c(x, y, \delta)^{-}, c(x, y, \delta)^{+}]$, reflects the strength of criteria in favor of xSy and depends on $C(xS_\delta y)$ and $D(xS_\delta y)$; similarly, $d_i(x, y)$ denotes the credibility index of the statement "xSy is vetoed by $g_i \in D(xS_\delta y)$" (cf. [17]).

The credibility of the assertion "x is at least as good as y" is calculated in two steps. First, a set of marginal credibility indices is defined for a set of attainable values δ as the conjunction of three predicates: (i) "the considered δ-concordance coalition has a sufficiently high credibility degree"; (ii) "the coalition is sufficiently strong"; and (iii) "the statement 'there is g_i that vetoes xSy' is not credible". Second, the marginal credibility indices are merged to produce the final credibility of xSy. For a given δ, the credibility index of the outranking for the δ-concordance coalition is defined as $\sigma_\delta(x, y) = \min\{\delta, P(c(x, y, \delta) \geq \boldsymbol{\lambda}), (1 - \max_{g_i \in D(xS_\delta y)} d_i(x, y))\}$. Let $\Gamma = \{\sigma_i(x, y) > 0; i = 1, 2, \ldots, N\}$ be the set of positive values of credibility indices of $xS_i y$ for the N criteria. Then, xSy with credibility index $\sigma(x, y) = \max\{\sigma_\delta\}(\delta \in \Gamma)$. If Γ is empty, then $\sigma(x, y) = 0$. Now, the DM can define a cutting threshold γ such that $xSy \Leftrightarrow \sigma(x, y) \geq \gamma . \sigma(x, y)$ can be exploited to rank the alternatives in A, assign them to ordered classes, or select the best of them. For example, using the following binary preference relations:

- $xSy \Leftrightarrow \sigma(x, y) \geq \gamma$: interval outranking relation;
- $xPy \Leftrightarrow \sigma(x, y) \geq \gamma$ and $\sigma(y, x) < \gamma$: interval preference relation;
- $xIy \Leftrightarrow \sigma(x, y) \geq \gamma$ and $\sigma(y, x) \geq \gamma$: interval indifference relation;
- $xRy \Leftrightarrow \sigma(x, y) < \gamma$ and $\sigma(y, x) < \gamma$: interval no-comparable relation.

Finally, a new dominance concept of action x over action y, denoted by $xD(\alpha)y$ ($\alpha \in \mathbb{R}$), is introduced by [17] as follows: $xD(\alpha)y \Leftrightarrow \min_{g_i \in G} P(\boldsymbol{g}_i(x) \geq \boldsymbol{g}_i(y)) \geq \alpha \geq 0.5$.

The interval-based outranking method, as well as the methods mentioned above, contain interesting properties and each one of them addresses some relevant characteristic of the project portfolio selection problem. An overall outranking-based method for Problem (1) must be able to build a representation of the DM's preferences and should be able to handle: (i) imperfect information through interval-based and pseudo-criteria; (ii) interacting criteria; (iii) a hierarchy of criteria; (iv) a ranking of projects or portfolios at the level of any non-elementary criterion; (v) a sorting of projects or portfolios also at the level of any non-elementary criterion; (vi) the different types of constraints that the DM desires to impose. Let us now provide an outline of such an overall method.

4.4 Using Evolutionary Algorithms to Optimize Portfolios

Solving the multi-objective optimization problem means finding the best compromise solution for the DM. This concept is ill-defined due to the conflicting nature of the objective functions. The best compromise concept is closely related to multi-criteria preferences. Balderas et al. [5] introduced a way to identify the best compromise solution to a multi-objective optimization problem (MOP). This is based on the next definition:

Definition 1 Let O be the image of the feasible region of the MOP. $x^* \in O$ is the best compromise solution to the problem only if it fulfills two conditions:

1. $\neg\exists y \in O : yPx^*$
2. x^*Sy for many solutions y such that $not(yPx^*)$.

The following concept was firstly introduced by Fernandez et al. [20], but in the context of classical ELECTRE methods. Below, it is adapted to the interval framework of Sect. 4.3.

Definition 2 Suppose that the DM is compatible with the interval outranking model of Sect. 4.3. Let O be the image of the feasible region of the MOP. $x \in O$ is called a non-strictly outranked portfolio if and only if there is no $y \in O$ such that yPx or $yD(\alpha)x$.

For a given $(\alpha, \lambda, \gamma)$, the set of non-strictly outranked solutions compose the non-strictly outranked frontier (NSF).

Remark 1

1. With appropriate values of γ, α, and λ, a non-strictly outranked solution from Definition 2 fulfills the first necessary condition to be the best compromise (Condition 1 of Definition 1).
2. The fulfillment of the second necessary condition (Condition 2 of Definition 1) is analyzed on the non-strictly outranked frontier, using the *outranking strength* measure. This measure is described as $\text{OS}(x) = card\{a_i \in NSF \text{ such that } xSa_i\}$, where NSF denotes the non-strictly outranked frontier.
3. More than one solution can satisfy Definition 1. The solution selected as the final best compromise may be the one with the highest value of outranking strength. Alternatively, in project portfolio optimization, the number of supported projects could be used as a complementary criterion to select the best compromise.

The MOP can be solved through the following steps:

- The DM (perhaps helped by a decision analyst) sets her/his model's parameters according to the interval outranking's characteristics.
- An approximation to the non-strictly outranked frontier is identified by an optimization algorithm; the set of solutions which fulfill Definition 1.1 is thus found.
- The DM chooses the best compromise solution according to Remark 1.

Balderas et al. [5] proposed the I-NOSGA (Interval Non-Outranked Sorting Genetic Algorithm) method to find the best compromise using an interval outranking model. The I-NOSGA method has as sources of inspiration the well-known NSGA-II [14], and NOSGA (cf. [20–22]). The I-NOSGA is an interval extension of NOSGA that identifies non-strictly outranked solutions in optimization problems. The main feature of I-NOSGA is its capacity to make selective pressure in the presence of uncertainty and incorporating the decision maker's preferences. I-NOSGA handles imperfect information through interval objective functions and constraints in optimization problems. Also, it integrates preferences with the use of non-strictly outranked fronts in the search process, rather than non-dominated ones as in NSGA-II. Finally, it combines the previous elements with the net strength score to exert selective pressure in the constructive process.

Solutions in I-NOSGA are encoded as a binary array of a size equivalent to the number of applicant projects, In this array, the ith element represents one gene whose information tells whether the ith project forms part of the portfolio. The genetic operators involved in the design of the algorithm are binary tournament, one-point crossover, and bit-flip mutation. Binary tournament selection chooses the best-ranked parent according to its front, randomly breaking ties.

After selection, the preference relation that determines the selective pressure in I-NOSGA depends weakly on the number of objective functions, it is expected to behave satisfactorily in problems with many objective functions. Balderas et al. [5] reported good results in a nine-objective project portfolio optimization problem with 100 applicant social-impact projects, with several segmentation constraints and imprecise information on project impacts and requirements, resource availability, and parameters of the decision model. I-NOSGA was successfully used in solving portfolio optimization of new product development projects under imprecise time-related dependencies and constraints [19].

Given its capacity to handle many objectives and imperfect information on criterion scores, model's parameters, requirement and availability of resources, I-NOSGA is a candidate to be chosen for solving Problem (1). However, I-NOSGA is not free of criticisms due to the following points:

- A high number of incomparability situations can be expected when dealing with a sufficiently large number of objectives, say more than six or seven; therefore, the selective pressure toward the true non-strictly outranked frontier could be weak in this scenario.
- I-NOSGA is sensitive to the selection of parameters to set the outranking and preference relation, particularly α, λ, and γ.
- In the experiments in [5], many solutions were found in the known non-strictly outranked frontier. The net strength scores of these solutions are very sensitive to the setting of λ and γ.

To overcome the above criticisms, a plausible way to increase the selective pressure within the interval outranking framework can be proposed. It consists of exploiting the truth degree of the predicate "x is preferred to y", which is the conjunction of "x is at least as good as y" and "y is not at least as good as x". Let $D_P(x, y)$ be this truth degree, then $DP(x, y) = \sigma(x, y) \wedge (1 - \sigma(y, x))$.

Let A be a set of solutions (e.g., the population in I-NOSGA), the truth degree that x is preferred to all $y \in A$, denoted as $P(x, A)$, is given by:

$$P(x, A) = \forall y_i \in A : DP(x, y_i) \Leftrightarrow DP(x, y_1) \wedge DP(x, y_2) \wedge \ldots \quad (3)$$

The truth degree of the conjunction can be calculated using the geometric mean as a logical operator (see [2] for the definition of fuzzy logic based on the geometric mean). Andrade et al. [16] proved a rank equivalence between Archimedean and compensatory operators. Particularly, the rank order derived from the geometric mean is the same as the one suggested by the product norm.

The best solution x^* must maximize $P(x^*, A)$, but it must also fulfill $\neg\exists y \in A : DP(y, x^*) > 0.5$; therefore, a single objective optimization technique can be used. The last condition is a non-strict outranked requirement in the sense of Definition 1.1. The maximization of $P(x, A)$ gives a sufficient condition in the sense of Definition 1.2.

5 Conclusions and Future Work

The characteristics of project portfolio optimization described along the chapter and formalized in Problem (1) show the necessity of metaheuristics to address this problem. The example discussed in Sect. 3.2 illustrates an important remark: even when the DM apparently has a single objective (for instance, maximizing the social components of the Human Development Index), he/she requires a complex multi-objective formulation to generate good portfolios. Evolutionary algorithms are commonly used in this environment, but so far, there are no published works that comprehensively deal with the stated problem. This work outlines an approach based on evolutionary algorithms that can handle the following situations: (i) portfolio selection problems under conditions of imprecision, uncertainty, ill-definition, and arbitrariness; (ii) decision makers made up of sets of individuals with conflicting points of view; (iii) many (even tens of thousands of) project proposals; (iv) interaction effects between projects; (v) segmentation constraints; (vi) conformity constraints and conformity objectives (related to the decision maker's agreement with the selected portfolio); (vii) many-objective functions.

The evolutionary approach should be able to deal with: (i) parameters defined as imprecise values; (ii) problems with many objective functions; (iii) objectives assessed with qualitative and ordinal information; (iv) exploit the preferential information provided by the decision maker seeking to lighten both computational and cognitive effort (e.g., performing a selective pressure towards the zone of the Pareto front where the most preferred portfolios are found). Some of the possible future research lines are the following:

- Extending the proposed approach to the assessment of potential solutions to project level.
- Performing numerical experiments to evaluate the approach's performance.
- Establishing a procedure to obtain the parameter values that better represent the DM's preferences.
- Finding the largest number of instances that can be solved with acceptable performances.
- Assessing the robustness of the approach.
- Applying the approach on an extensive study case.

Acknowledgements Carlos A. Coello Coello acknowledges support from CONACyT project no. 2016-01-1920 (*Investigación en Fronteras de la Ciencia 2016*) and from a project from the 2018 SEP-Cinvestav Fund (application no. 4).

References

1. de Almeida, A.T., Duarte, M.D.: A multi-criteria decision model for selecting Project Portfolio with consideration being given to a new concept for synergies. Pesqui. Oper. **31**(2), 301–318 (2011). https://doi.org/10.1590/S0101-74382011000200006. www.scielo.br/pope
2. Andrade, R.A.E., Fernández, E., González, E.: Compensatory fuzzy logic: a frame for reasoning and modeling preference knowledge in intelligent systems. Studies in Computational Intelligence, vol. 537, pp. 3–23 (2014). https://doi.org/10.1007/978-3-642-53737-0_1. https://link.springer.com/chapter/10.1007/978-3-642-53737-0_1
3. Arratia, M.N., López, I.F., Schaeffer, S.E., Cruz-Reyes, L.: Static R&D project portfolio selection in public organizations. Decis. Support Syst. **84**, 53–63 (2016). https://doi.org/10.1016/j.dss.2016.01.006
4. Baker, E., Bosetti, V., Salo, A.: Robust portfolio decision analysis: an application to the energy research and development portfolio problem. Eur. J. Oper. Res. **284**(3), 1107–1120 (2020). https://doi.org/10.1016/j.ejor.2020.01.038
5. Balderas, F., Fernandez, E., Gomez-Santillan, C., Rangel-Valdez, N., Cruz, L.: An interval-based approach for evolutionary multi-objective optimization of project portfolios. Int. J. Inf. Technol. Decis. Mak. **18**(4), 1317–1358 (2019). https://doi.org/10.1142/S021962201950024X
6. Barbati, M., Greco, S., Kadziński, M., Słowiński, R.: Optimization of multiple satisfaction levels in portfolio decision analysis. Omega (United Kingdom) **78**, 192–204 (2018). https://doi.org/10.1016/j.omega.2017.06.013
7. Bouyssou, D., Marchant, T., Perny, P., Pirlot, M., Tsoukás, A., Vincke, P.: Evaluation And Decision Models: A Critical Perspective, 1st edn. Springer, New York (2000). https://doi.org/10.1007/978-1-4615-1593-7
8. Carazo, A.F., Gómez, T., Molina, J., Hernández-Díaz, A.G., Guerrero, F.M., Caballero, R.: Solving a comprehensive model for multiobjective project portfolio selection. Comput. Oper. Res. **37**(4), 630–639 (2010). https://doi.org/10.1016/j.cor.2009.06.012
9. Chassein, A., Goerigk, M., Kasperski, A., Zieliński, P.: On recoverable and two-stage robust selection problems with budgeted uncertainty. Eur. J. Oper. Res. **265**(2), 423–436 (2018). https://doi.org/10.1016/j.ejor.2017.08.013
10. Cooper, R., Edgett, S., Kleinschmidt, E.: Portfolio management for new product development: results of an industry practices study. R D Manag. **31**(4), 361–380 (2001). https://doi.org/10.1111/1467-9310.00225
11. Corrente, S., Figueira, J.R., Greco, S., Słowiński, R.: A robust ranking method extending ELECTRE III to hierarchy of interacting criteria, imprecise weights and stochastic analysis. Omega (United Kingdom) **73**, 1–17 (2017). https://doi.org/10.1016/j.omega.2016.11.008
12. Corrente, S., Greco, S., SŁowiński, R.: Multiple criteria hierarchy process with ELECTRE and PROMETHEE. Omega (United Kingdom) **41**(5), 820–846 (2013). https://doi.org/10.1016/j.omega.2012.10.009
13. Corrente, S., Greco, S., Słowiński, R.: Multiple criteria hierarchy process for ELECTRE Tri methods. Eur. J. Oper. Res. **252**(1), 191–203 (2016). https://doi.org/10.1016/j.ejor.2015.12.053
14. Deb, K., Pratap, A., Agarwal, S., Meyarivan, T.: A fast and elitist multiobjective genetic algorithm: NSGA-II. IEEE Trans. Evol. Comput. **6**(2), 182–197 (2002). https://doi.org/10.1109/4235.996017
15. Doerner, K.F., Gutjahr, W.J., Hartl, R.F., Strauss, C., Stummer, C.: Pareto ant colony optimization with ILP preprocessing in multiobjective project portfolio selection. Eur. J. Oper. Res. **171**(3), 830–841 (2006). https://doi.org/10.1016/j.ejor.2004.09.009

16. Espin-Andrade, R.A., Caballero, E.G., Pedrycz, W., Fernández González, E.R.: Archimedean-compensatory fuzzy logic systems. Int. J. Comput. Intell. Syst. **8**(Supplement 2), 54–62 (2015). https://doi.org/10.1080/18756891.2015.1129591. www.uas.edu.mx
17. Fernández, E., Figueira, J.R., Navarro, J.: An interval extension of the outranking approach and its application to multiple-criteria ordinal classification. Omega (United Kingdom) **84**, 189–198 (2019). https://doi.org/10.1016/j.omega.2018.05.003
18. Fernández, E., Figueira, J.R., Navarro, J.: Interval-based extensions of two outranking methods for multi-criteria ordinal classification. Omega (United Kingdom) **95**, 102065 (2020). https://doi.org/10.1016/j.omega.2019.05.001
19. Fernandez, E., Gomez-Santillan, C., Rangel-Valdez, N., Cruz-Reyes, L., Balderas, F.: An interval-based evolutionary approach to portfolio optimization of new product development projects. Math. Probl. Eng. **2019** (2019). https://doi.org/10.1155/2019/4065424
20. Fernandez, E., Lopez, E., Bernal, S., Coello Coello, C.A., Navarro, J.: Evolutionary multiobjective optimization using an outranking-based dominance generalization. Comput. Oper. Res. **37**(2), 390–395 (2010). https://doi.org/10.1016/j.cor.2009.06.004
21. Fernandez, E., Lopez, E., Lopez, F., Coello Coello, C.A.: Increasing selective pressure towards the best compromise in evolutionary multiobjective optimization: the extended NOSGA method. Inf. Sci. **181**(1), 44–56 (2011). https://doi.org/10.1016/j.ins.2010.09.007
22. Fernandez, E., Lopez, E., Mazcorro, G., Olmedo, R., Coello Coello, C.A.: Application of the non-outranked sorting genetic algorithm to public project portfolio selection. Inf. Sci. **228**, 131–149 (2013). https://doi.org/10.1016/j.ins.2012.11.018
23. Fernández González, E.: Análisis de la decisión para evaluar y formar carteras de proyectos de I&D: un enfoque Crítico Y Propositivo. Gestión y Política Pública **XII**(1), 43–79 (2003). https://www.redalyc.org/articulo.oa?id=13312102
24. Figueira, J.R., Greco, S., Roy, B.: ELECTRE methods with interaction between criteria: an extension of the concordance index. Eur. J. Oper. Res. **199**(2), 478–495 (2009). https://doi.org/10.1016/j.ejor.2008.11.025
25. Fox, G.E., Baker, N.R., Bryant, J.L.: Economic models for R and D project selection in the presence of project interactions. Manag. Sci. **30**(7), 890–902 (1984). https://doi.org/10.1287/mnsc.30.7.890
26. Gatzert, N., Martin, A., Schmidt, M., Seith, B., Vogl, N.: Portfolio optimization with irreversible long-term investments in renewable energy under policy risk: a mixed-integer multistage stochastic model and a moving-horizon approach. Eur. J. Oper. Res. **290**(2), 734–748 (2021). https://doi.org/10.1016/j.ejor.2020.08.033
27. Ghasemzadeh, F., Archer, N., Lyogun, P.: A zero-one model for project portfolio selection and scheduling. J. Oper. Res. Soc. **50**(7), 745–755 (1999). https://doi.org/10.1057/palgrave.jors.2600767
28. Gunasekaran, A., Kirytopoulos, K., Leopoulos, V., Mavrotas, G., Voulgaridou, D.: Multiple sourcing strategies and order allocation: an ANPAUGMECON meta model. Supply Chain. Manag.: Int. J. **15**(4), 263–276 (2010). https://doi.org/10.1108/13598541011054643
29. Hao, P., Guo, J.P., Chen, Y., Xie, B.C.: Does a combined strategy outperform independent policies? Impact of incentive policies on renewable power generation. Omega (United Kingdom) **97**, 102100 (2020). https://doi.org/10.1016/j.omega.2019.08.007
30. Henriksen, A.D., Traynor, A.J.: A practical R&D project-selection scoring tool. IEEE Trans. Eng. Manage. **46**(2), 158–170 (1999). https://doi.org/10.1109/17.759144
31. Klapka, J., Piňos, P.: Decision support system for multicriterial R&D and information systems projects selection. Eur. J. Oper. Res. **140**(2), 434–446 (2002). https://doi.org/10.1016/S0377-2217(02)00081-4
32. Korotin, V., Popov, V., Tolokonsky, A., Islamov, R., Ulchenkov, A.: A multi-criteria approach to selecting an optimal portfolio of refinery upgrade projects under margin and tax regime uncertainty. Omega (United Kingdom) **72**, 50–58 (2017). https://doi.org/10.1016/j.omega.2016.11.003
33. Korotkov, V., Wu, D.: Evaluating the quality of solutions in project portfolio selection. Omega (United Kingdom) **91**, 102029 (2020). https://doi.org/10.1016/j.omega.2019.01.007

34. Korotkov, V., Wu, D.: Benchmarking project portfolios using optimality thresholds. Omega (United Kingdom) **99**, 102166 (2021). https://doi.org/10.1016/j.omega.2019.102166
35. Li, X., Huang, Y.H., Fang, S.C., Zhang, Y.: An alternative efficient representation for the project portfolio selection problem. Eur. J. Oper. Res. **281**(1), 100–113 (2020). https://doi.org/10.1016/j.ejor.2019.08.022
36. Liberatore, M.J., Titus, G.J.: Practice of management science in R&D project management. Manag. Sci. **29**(8), 962–974 (1983). https://doi.org/10.1287/mnsc.29.8.962
37. Liesiö, J., Mild, P., Salo, A.: Preference programming for robust portfolio modeling and project selection. Eur. J. Oper. Res. **181**(3), 1488–1505 (2007). https://doi.org/10.1016/j.ejor.2005.12.041
38. Litvinchev, I.S., López, F., Alvarez, A., Fernández, E.: Large-scale public R&D portfolio selection by maximizing a biobjective impact measure. IEEE Trans. Syst. Man Cybern. Part A Syst. Hum. **40**(3), 572–582 (2010). https://doi.org/10.1109/TSMCA.2010.2041228
39. Martinsuo, M., Geraldi, J.: Management of project portfolios: relationships of project portfolios with their contexts. Int. J. Project Manage. **38**(7), 441–453 (2020). https://doi.org/10.1016/j.ijproman.2020.02.002
40. Mavrotas, G., Diakoulaki, D., Caloghirou, Y.: Project prioritization under policy restrictions. A combination of MCDA with 0-1 programming. Eur. J. Oper. Res. **171**(1), 296–308 (2006). https://doi.org/10.1016/j.ejor.2004.07.069
41. Mavrotas, G., Diakoulaki, D., Kourentzis, A.: Selection among ranked projects under segmentation, policy and logical constraints. Eur. J. Oper. Res. **187**(1), 177–192 (2008). https://doi.org/10.1016/j.ejor.2007.03.010
42. Mavrotas, G., Makryvelios, E.: Combining multiple criteria analysis, mathematical programming and monte carlo simulation to tackle uncertainty in research and development project portfolio selection: A case study from greece. Eur. J. Oper. Res. **291**(2), 794–806 (2021)
43. Mavrotas, G., Pechak, O.: Combining mathematical programming and monte carlo simulation to deal with uncertainty in energy project portfolio selection. Green Energy Technol. **129**, 333–356 (2013). https://doi.org/10.1007/978-1-4471-5143-2_16
44. Medaglia, A.L., Graves, S.B., Ringuest, J.L.: A multiobjective evolutionary approach for linearly constrained project selection under uncertainty. Eur. J. Oper. Res. **179**(3), 869–894 (2007). https://doi.org/10.1016/j.ejor.2005.03.068
45. Moore, R.E.: Interval arithmetic and automatic error analysis in digital computing. Ph. D. dissertation,, Stanford University, Stanford, CA, United States (1963). https://dl.acm.org/doi/10.5555/904631
46. Morton, A., Keisler, J.M., Salo, A.: Multicriteria portfolio decision analysis for project selection. Int. Ser. Oper. Res. Manag. Sci. **233**, 1269–1298 (2016). https://doi.org/10.1007/978-1-4939-3094-4_28
47. Mukherjee, K., Bera, A.: Application of goal programming in project selection decision - a case study from the Indian coal mining industry. Eur. J. Oper. Res. **82**(1), 18–25 (1995). https://doi.org/10.1016/0377-2217(94)00197-K
48. Programme, U.N.D.: Human Development Report 1997. Oxford University, Oxford, Technical report (1997)
49. Relich, M., Pawlewski, P.: A fuzzy weighted average approach for selecting portfolio of new product development projects. Neurocomputing **231**, 19–27 (2017). https://doi.org/10.1016/j.neucom.2016.05.104
50. Ritzenhofen, I., Birge, J.R., Spinler, S.: The structural impact of renewable portfolio standards and feed-in tariffs on electricity markets. Eur. J. Oper. Res. **255**(1), 224–242 (2016). https://doi.org/10.1016/j.ejor.2016.04.061
51. Roy, B.: The outranking approach and the foundations of electre methods. Theory Decis. **31**(1), 49–73 (1991). https://doi.org/10.1007/BF00134132
52. Roy, B., Figueira, J., Almeida-Dias, J.: Discriminating thresholds as a tool to cope with imperfect knowledge in multiple criteria decision aiding: theoretical results and practical issues. Omega **43**, 9–20 (2014)

53. Salo, A., Keisler, J., Morton, A. (eds.): Portfolio Decision Analysis. International Series in Operations Research & Management Science, vol. 162. Springer, New York (2011). https://doi.org/10.1007/978-1-4419-9943-6
54. Stummer, C., Heidenberger, K.: Interactive R&D portfolio analysis with project interdependencies and time profiles of multiple objectives. IEEE Trans. Eng. Manage. **50**(2), 175–183 (2003). https://doi.org/10.1109/TEM.2003.810819
55. Szilágyi, I., Sebestyén, Z., Tóth, T.: Project ranking in petroleum exploration. Eng. Econ. **65**(1), 66–87 (2020). https://doi.org/10.1080/0013791X.2019.1593570
56. Tervonen, T., Liesiö, J., Salo, A.: Modeling project preferences in multiattribute portfolio decision analysis. Eur. J. Oper. Res. **263**(1), 225–239 (2017). https://doi.org/10.1016/j.ejor.2017.04.051
57. Toppila, A., Salo, A.: Binary decision diagrams for generating and storing non-dominated project portfolios with interval-valued project scores. Eur. J. Oper. Res. **260**(1), 244–254 (2017). https://doi.org/10.1016/j.ejor.2016.12.019
58. Vetschera, R., De Almeida, A.T.: A PROMETHEE-based approach to portfolio selection problems. Comput. Oper. Res. **39**(5), 1010–1020 (2012). https://doi.org/10.1016/j.cor.2011.06.019
59. Yunes, T.H., Napolitano, D., Scheller-Wolf, A., Tayur, S.: Building efficient product portfolios at John Deere and Company. Oper. Res. **55**(4), 615–629 (2007). https://doi.org/10.1287/opre.1070.0405
60. Zanakis, S.H., Mandakovic, T., Gupta, S.K., Sahay, S., Hong, S.: A review of program evaluation and fund allocation methods within the service and government sectors. Socioecon. Plann. Sci. **29**(1), 59–79 (1995). https://doi.org/10.1016/0038-0121(95)98607-W

A New Model for the Project Portfolio Selection and Scheduling Problem with Defence Capability Options

Kyle Robert Harrison, Saber M. Elsayed, Ivan L. Garanovich, Terence Weir, Sharon G. Boswell, and Ruhul Amin Sarker

Abstract This chapter proposes a novel optimization model for the project portfolio selection and scheduling problem (PPSSP), as part of the future force design (FFD) process, in the context of the Australian Defence Force (ADF) capability development. The core objective of the problem addressed in this study is to maximize the total portfolio value attained by the selection and scheduling of a set of projects, grouped in various subsets referred to as capability options (COs), while adhering to budgetary, scheduling, and operational constraints. While many of the independent aspects of this problem have been considered in the literature, no existing problem formulation adequately captures all aspects. To solve the proposed model, a custom-built heuristic approach is first developed. Three meta-heuristic approaches are examined to improve upon the heuristic solutions. Experiments are conducted over a set of 108 synthetic problem instances, depicting a wide variety of problem characteristics.

K. R. Harrison (✉) · S. M. Elsayed · R. A. Sarker
University of New South Wales Canberra at the Australian Defence Force Academy, Canberra, Australia
e-mail: kyle.harrison@unsw.edu.au

S. M. Elsayed
e-mail: s.elsayed@unsw.edu.au

R. A. Sarker
e-mail: r.sarker@adfa.edu.au

I. L. Garanovich · T. Weir · S. G. Boswell
Joint Operations and Analysis Division, Defence Science and Technology Group, Department of Defence, Canberra, Australia
e-mail: ivan.garanovich@dst.defence.gov.au

T. Weir
e-mail: terence.weir@dst.defence.gov.au

S. G. Boswell
e-mail: sharon.boswell@dst.defence.gov.au

K. R. Harrison et al., *Evolutionary and Memetic Computing for Project Portfolio Selection and Scheduling*, Adaptation, Learning, and Optimization 26,
https://doi.org/10.1007/978-3-030-88315-7_5

1 Introduction

In the Australian Defence Force (ADF), future force design (FFD) is a complex task that assists with the planning of critical defense capability investment decisions shaping a future defense force. The main deliverable of the FFD process is the selection of a portfolio of projects from a large number of strategic investment or divestment options, each of which has an effect on the future capabilities of the ADF. In this context, a capability refers to the ability to achieve an operational effect through integral components such as doctrine, training, and equipment development [4]. The portfolio of projects is known as the Integrated Investment Program (IIP) [6] and can consist of both investment and divestment projects, i.e., projects that deliver additional capabilities and have a positive net cost or projects that release previously allocated funds at the expense of decreasing the delivery of future capabilities. The number of available projects is often quite large, leading to a very challenging optimization problem. The high-level objective is to maximize the total portfolio value, i.e., maximize the delivery of capabilities, while satisfying the various, complex operational constraints. More specifically, the primary objective can be stated as the development of a robust set of low-level capabilities that can adequately respond to any requirements that may arise in the future.

Previous studies have examined this problem using a largely traditional formulation of the project portfolio selection and scheduling problem (PPSSP) [7, 8, 11, 12, 17, 19, 20, 24, 25, 29, 31], whereby the primary selection is done at the project level. For a recent survey on portfolio optimization in the defense sector, the reader is referred to [16]. In these previous problem formulations, the requirement is to select a set of projects and, in some cases, their respective start times subject to various operational constraints. The value of the portfolio is then dictated by the sum of the values associated with each of the projects, possibly including synergies between various projects. As identified by [16], one of the major weaknesses of previous approaches to PPSSP is the lack of time-varying valuation of the implementations. That is, many previous studies do not consider that the value of a project may be greater if it is initiated/completed earlier. Furthermore, many of the existing approaches in the literature have made use of exact solvers [16], representing a disconnect between the scale of problems faced in real-world defense applications and the scholarly literature surrounding the PPSSP. While previous approaches are suitable for many PPSSP problems, they do not match the requirements of the modern FFD process undertaken by the ADF.

Due to the extremely complex nature of the modern FFD process, which needs to capture and address numerous strategic, operational, and tactical interdependencies between various defense capabilities and capability projects, a new approach to the PPSSP is required to meet this challenge in a practical way. In the proposed problem formulation, the primary unit of selection is known as a Capability option (CO), which is a subset of individual projects that, when implemented together, provide an operational capability. It is noteworthy that the subset of projects in each CO is not necessarily unique, i.e., a project may be included in several COs, and that the projects can correspond to either investments (i.e., positive cost and positive value) or divestments (i.e., negative costs and negative values). The mix of investment and divestment options is a feature that is not often considered in the literature. A CO can only be selected as a whole unit such that all the constituent projects must be selected if the CO is implemented. Thus, there is no longer selection at the independent project level in this new formulation.

The COs are then grouped into independent *families*, such that one CO must be selected from each family. The "value" associated with a CO is determined by the earliest time period in which all individuals in the CO can reach an initial level of their operating capability. Note that, a project can appear in multiple COs, but can only be included in the portfolio once. Hence, if a project belongs to multiple COs that are selected for implementation, the cost of the project must only be counted once. However, the net effect of a project on multiple COs is not diminished in this scenario. Thus, there is an inherent benefit in selecting projects that belong to multiple COs. Furthermore, various operational constraints are present, such as budget considerations, scheduling milestones, and value synergies between various COs and/or individual projects. While similar binary selection problems have been examined in the literature, no single existing problem formulation adequately addresses all aspects. Most notably, the primary unit of selection in the proposed problem is the CO, which has the potential to contain overlapping sets of projects. Furthermore, the proposed problem encompasses both investment and divestment options whereas most existing selection problems make an explicit assumption that items will have a positive value and weight/cost. Hence, existing heuristics are not appropriate and must be adapted for use on the proposed problem formulation.

As a preliminary approach to solving this problem, a custom-built heuristic is developed. This heuristic approach is inspired by some of the existing heuristic techniques proposed for closely related problems in the literature [21, 32]. At a high level, the heuristic process first selects a set of COs that correspond to divestment options, thus providing an increase to the available budget and a decrease to the overall portfolio value. Next, a greedy selection process is used to select a set of COs that correspond to investment options, thereby diminishing the available budget while increasing the portfolio value. Finally, a local search process, based on swapping of COs among each family, is used to refine the initial greedy solution. The objective is to provide a net positive value through this process while balancing the cost associated with the selected divestment and investment options. While this heuristic approach is expected to provide reasonable initial solutions, evolutionary meta-heuristics are subsequently employed for refinement. A set of 108 problem instances for the proposed model are generated and solved using the heuristic and meta-heuristic approaches. The effect of various problem characteristics, such as the number of families, projects, and the proportion of divestment options, is examined.

The remainder of this chapter is structured as follows. Section 2 provides the reader with necessary background information. Section 3 details the proposed problem formulation. Section 4 proposes an initial heuristic approach to address the proposed problem. The experimental design is given in Sect. 5, the results of which are presented and discussed in Sect. 6. Finally, concluding remarks are given in Sect. 7.

2 Background

This section provides the reader with the background necessary for the remainder of this chapter. First, this section provides an overview of knapsack problems, which are inherently related to the problem examined in this study, then provides background on evolutionary meta-heuristic approaches.

2.1 The Knapsack Problem

The most common formulation of the knapsack problem is the 0–1 (or binary) knapsack problem, with the definition as follows. Given a set of n items, each with an associated profit p_i and unit weight w_i, the objective is to maximize the profit of selected items while adhering to the total weight capacity C. Formally, the traditional knapsack problem can be defined by

$$\max \sum_{i=1}^{n} p_i x_i \tag{1a}$$

subject to

$$\sum_{i=1}^{n} w_i x_i \leq C \tag{1b}$$

$$x_i \in \{0, 1\}, \quad \forall i \in 1, \ldots, n \tag{1c}$$

such that x_i takes the value 1 if item i is selected for inclusion in the knapsack.

2.1.1 Multidimensional Knapsack Problem

One of the simplest extensions of the knapsack problem is the multidimensional knapsack problem (MKP) [27], which replaces the scalar item weights and knapsack capacity with vectors of length m. The problem formulation remains largely unchanged, with the exception of the capacity constraint reformulation to ensure that the capacity defined in each dimension is respected. Formally, this problem can be defined by

$$\max \sum_{i=1}^{n} p_i x_i \tag{2a}$$

subject to

$$\sum_{i=1}^{n} w_{ij} x_i \leq C_j, \quad \forall j = 1, \ldots, m \tag{2b}$$

$$x_i \in \{0, 1\}, \quad \forall i \in 1, \ldots, n. \tag{2c}$$

By defining the weight and capacity vectors to be scalars, i.e., vectors of length 1, the MKP reduces to the traditional knapsack problem.

2.1.2 Multiple-Choice Knapsack Problem

The multiple-choice knapsack problem (MCKP) [23] was proposed as a generalization of the traditional knapsack problem whereby the set of items is divided into distinct classes such that only one item from each class can be selected. Given a set of m mutually disjoint classes $(N_1, \ldots, N_m)$ of items, where each item has a profit p_{ij} and weight w_{ij}, and a total capacity of C, the objective is to maximize the profit while selecting exactly one item from each class without exceeding the capacity. Using a binary variable x_{ij}, defined to take the value 1 if item j is selected from class N_i, the MCKP can be formally defined by

$$\max \sum_{i=1}^{m} \sum_{j \in N_i} p_{ij} x_{ij} \tag{3a}$$

subject to

$$\sum_{i=1}^{m} \sum_{j \in N_i} w_{ij} x_{ij} \leq C \tag{3b}$$

$$\sum_{j \in N_i} x_{ij} = 1, \quad \forall i \in 1, \ldots, m \tag{3c}$$

$$x_{ij} \in \{0, 1\}, \quad \forall i \in 1, \ldots, n \text{ and } \forall j \in N_i. \tag{3d}$$

A special formulation can be used to reduce the MCKP to the traditional knapsack formulation. Specifically, each class must be defined to contain exactly two items, namely one item and an "empty" item with no weight or profit (i.e., $p_{ij} = w_{ij} = 0$), which is an analog for not selecting a particular item in the traditional formulation.

A further extension of the MCKP also incorporates the vector weights and capacity from the MKP, giving rise to the multiple-choice multidimensional knapsack problem (MMKP) [21].

2.1.3 Set-Union Knapsack Problem

The Set-union knapsack problem (SUKP) [13] was proposed as another generalized model of the knapsack problem. The general concept is that the selection process works on a collection of items, which are in turn comprised of a subset of the available elements, such that the total weight is dictated by the union of the selected elements. That is, the weight is only counted once for each element, regardless of the number of times it is included in a selected item. The definition of the SUKP is as follows. Given a set U of n elements, where each element j has a weight $w_j > 0$, a set V of m items, where each item i is a subset of elements $U_i \subseteq U$ with an associated profit $p_i > 0$, and a knapsack capacity C, the objective is to determine a set of items $S \subseteq V$ that maximizes the total profit associated with S such that the total weight of elements contained in S does not exceed C. Formally, the SUKP can be formulated as

$$\max \sum_{i \in S} p_i \tag{4a}$$

subject to:

$$\sum_{j \cup_{i \in S} U_i} w_j \leq C, S \subseteq V. \tag{4b}$$

Note that, by defining each element subset to contain exactly one distinct item, the formulation of Eq. (4) reduces to the traditional knapsack problem.

2.2 Evolutionary Meta-Heuristic Approaches

This section presents a brief introduction to the evolutionary meta-heuristic approaches that are used in this study.

2.2.1 Genetic Algorithms and the Biased Random-Key Genetic Algorithm

The Genetic Algorithm (GA) [22] is a population-based meta-heuristic inspired by the concept of Darwinian evolution. A population of candidate solutions, referred to as chromosomes, are evolved using selection, recombination, and mutation operators such that the selection mechanism provides selection pressure on the better-fit solutions, thereby improving the overall fitness of the population over time.

At each generation, candidate solutions are evaluated according to the objective function, and then used to generate a new population for the next generation. If desired, a proportion of the best-fit solutions are first inserted directly into the next generation via a process referred to as *elitism*. Elitism ensures that the best-known genetic material is preserved at each generation, thereby preventing a degradation of the objective fitness over time. A mating pool is formed by selecting chromosomes, generally with a bias towards those with better fitness values. Members of the mating pool are recombined to form new chromosomes, referred to as offspring, via a crossover operator. The generated offspring may then undergo a mutation operation, depending on the mutation rate parameter, before being inserted into the new population. In general, the mutation operator performs a slight perturbation on a single chromosome, with the intention of introducing new genetic material into the population. This entire process repeats until the desired termination criteria are met.

The Random-Key Genetic Algorithm (RKGA) [1] was proposed as a real-valued variant of the GA [22] to address combinatorial optimization problems. A chromosome in the RKGA consists of real-valued numbers in the range [0, 1], referred to as random keys, which are then decoded to the corresponding (discrete) solution for evaluation. For example, decoding can consist of sorting the keys (for a permutation problem) or using binning (for a discrete problem). The population is then sorted according to the objective fitness. An elite population is formed by selecting a proportion of the best fit chromosomes, which are then directly copied to the next generation as a form of elitism. Recombination is done by randomly selecting two chromosomes from the entire population and performing a uniform crossover operation. Mutation is done using an immigration strategy whereby the chromosome initialization procedure is used to generate new random chromosomes, which are directly inserted into the new population. With the exception of the decoding process, all steps in the RKGA are problem-independent.

The Biased Random-Key Genetic Algorithm (BRKGA) [14] is an extension of the RKGA whereby recombination is performed using one parent from the elite population and one parent from the non-elite population. Additionally, the crossover operation is biased towards one of the parents using a non-uniform selection probability, referred to as the *bias*. Typically, the bias would be in favor of the better fit parent, i.e., the parent from the elite population. Note that, if the *bias* is set to 0.5 (50%), then the biased crossover is equivalent to the uniform crossover from RKGA. In general, BRKGA has been shown to exhibit better performance than RKGA [14, 15].

2.3 Differential Evolution

Differential Evolution (DE) [30] is a popular evolutionary optimization technique that evolves a population of real-valued individuals using recombination, mutation, and selection. To generate new candidate solutions from the existing population, one or more *trial* positions are created through a recombination operator. If a trial position improves the objective fitness of the individual and remains within the feasible search domain, the trial position is accepted and replaces the current position of the individual. Otherwise, the trial position is discarded and the individual retains its current position. Similar to a GA, different operators can be employed in DE to tailor the evolutionary process to the current problem being considered. The choice of operators is usually denoted by DE/s/n/c where s is the selection operator, n is the number of trial positions to be generated, and c is the crossover operator.

The most common variant of DE is known as DE/rand/1/bin [26], whereby random selection is used, a single trial position is created, and crossover is performed using a binary operator. Creation of the trial position $\mathbf{t}$ for an individual $\mathbf{x}$ in dimension i using the DE/rand/1/bin strategy is done according to

$$t_i = \begin{cases} a_i + F(b_i - c_i) & \text{if } rand() < c_r \text{ or } i = RD \\ x_i & \text{otherwise} \end{cases}, \tag{5}$$

where $\mathbf{a}$, $\mathbf{b}$, and $\mathbf{c}$ are three randomly selected, distinct individuals, i.e., $\mathbf{a} \neq \mathbf{b} \neq \mathbf{c} \neq \mathbf{x}$, $F \in [0, 2]$ is the differential weight, $rand() \sim U(0, 1)$, RD is a uniform random integer from the range $[1, D]$ that is independently selected for each recombination operation; D is the number of problem dimensions; and $c_r \in [0, 1]$ is the crossover probability. Note that, the trial vector will always contain at least one dimension (RD) that was generated through recombination, ensuring that new genetic material is introduced.

3 Problem Formulation

In the problem formulation considered in this study, individual projects are partitioned into different groups, known as COs. Furthermore, each of the COs belongs to a family such that exactly one CO must be selected from each family. The COs, and as a result the families as well, can contain a combination of both investments and divestments. The objective of the problem is to maximize the delivery of capabilities by balancing investment and divestment options such that the net cost is minimized. Although the problem formulation is not unique, one possible formulation can be given as:

Notation
Subscripts

i	family
j	capability option
p	project
t	time period within the planning horizon

Sets

P	set of projects
O_i	set of COs in family i
F_{ij}	set of projects in family i and CO j
I	set of families

Variables

$$y_{ij} = \begin{cases} 1 & \text{if CO } j \text{ from family } i \text{ is selected} \\ 0 & \text{otherwise} \end{cases}$$

$$z_{pt} = \begin{cases} 1 & \text{if project } p \text{ is selected to start in time period } t \\ 0 & \text{otherwise} \end{cases}$$

Data and Parameters

d_p	duration of project p (in years)
te_p	time to initial capability delivery (i.e., effectiveness) for project p
EST_p	earliest allowable start time for project p
LST_p	latest allowable start time for project p
NP_{ij}	number of projects in CO j in family i
CP_{pt}	cost of project p in period t
BD_t	allowable budget deviation in time period t
E_{ij}	effective time of CO j in family i
VO_{ijt}	value of CO j in family i with effective time t

Constraints

Option selection: only one CO can be selected from each family

$$\sum_{j \in O_i} y_{ij} = 1, \quad \forall i \in I. \tag{6}$$

Project start time: every selected project must start within their earliest and latest start time windows

$$\sum_{t=EST_p}^{LST_p} z_{pt} \le 1, \quad \forall p \in P. \tag{7}$$

Linking projects with options: if a CO is selected, all projects within that CO must be selected

$$\sum_{p \in F_{ij}} \sum_{t=EST_p}^{LST_p} z_{pt} = NP_{ij} \times y_{ij}, \quad \forall i, j. \tag{8}$$

Identifying active projects: to calculate yearly project costs, the current period (k) must be differentiated from the start time of project p. If project p starts in time t and lasts d_p periods, then the execution time of p in period k is $k + 1 - t$. If $k + 1 - t \le 0$, then the project has not yet started whereas if $k + 1 - t > d_p$, the project has been completed. Thus, project p will be active in period t if

$$\sum_{t=k-d_p+1}^{k} z_{pt} = 1. \tag{9}$$

Yearly costs and budget deviation: this model considers both investments and divestments, with positive and negative costs, respectively. Divestments supply money while investments consume money. It is assumed that deviations from balancing the budget (i.e., a net expenditure of 0) are acceptable within a given limit BD_k, defined for each period k

$$\sum_{p=1}^{|P|} \sum_{t=k-d_p+1}^{k} z_{pt} \times CP_{pk} \leq BD_k, \quad \forall k. \tag{10}$$

Determining the effective time of a capability option is based on the latest effective time of all projects within the option. The effective time of a C O is taken as the earliest time period in which all constituent projects provide an initial capability. For a particular project, the time to deliver an initial capability is based on the period in which it starts and its time until effectiveness (te_p). If not selected, E_{ij} will be 0. The effective time can be calculated as

$$E_{ij} = \max_{p \in F_{ij}} \sum_{t=EST_p}^{LST_p} z_{pt}(t + te_p), \quad \forall i, j. \tag{11}$$

Objective

Value of selected options: each CO has an associated value, which depends on when the constituent projects can deliver their initial capabilities (i.e., E_{ij}), such that divestment options decrease the portfolio value whereas investment options increase the portfolio value. The total portfolio value is the sum of positive and negative values associated with the selected COs, and further depends on when all projects in a CO can provide an initial level o f their operating capabilities

$$\sum_{i=1}^{|I|} \sum_{j \in O_i} y_{ij} \times VO_{ijE_{ij}}. \tag{12}$$

Model

Given the objective and constraints above, the corresponding optimization model is formulated as:

$$\max \sum_{i=1}^{|I|} \sum_{j \in O_i} y_{ij} \times VO_{ijt} \tag{13a}$$

subject to:

$$\sum_{j \in O_i} y_{ij} = 1, \quad \forall i \in I \tag{13b}$$

$$\sum_{t=EST_p}^{LST_p} z_{pt} \leq 1, \quad \forall p \in P \tag{13c}$$

$$\sum_{t=EST_p}^{LST_p} \sum_{p \in F_{ij}} z_{pt} = NP_{ij} \times y_{ij}, \quad \forall i, j \tag{13d}$$

$$\sum_{t=k-d_p+1}^{k} z_{pt} = 1 \tag{13e}$$

$$\sum_{p=1}^{|P|} \sum_{t=k-d_p+1}^{k} z_{pt} \times CP_{pk} \leq BD_k, \quad \forall k \tag{13f}$$

$$y_{ij} \in \{0, 1\}, \quad \forall i, j \tag{13g}$$

$$z_{pt} \in \{0, 1\}, \quad \forall p, t \tag{13h}$$

3.1 Analysis of Problem Formulation

Assuming there are I families, an average of J options in each family, N_p independent projects, an average of P periods in the feasible starting window for each project, and T periods in the planning horizon, then the number of variables and constraints is as follows.

Regarding the number of variables

- y_{ij} has $I \times J$ variables
- z_{pt} has $N_p \times T$ variables,

for a total of $(I \times J) + (N_p \times T)$ variables.

Regarding the number of constraints,

- Equation (6) defines I constraints
- Equation (7) defines $N_p \times P$ constraints
- Equation (8) defines $I \times J$ constraints
- Equation (10) defines T constraints,

for a total of $I + (N_p \times P) + (I \times J) + T$ constraints.

As an example, consider a scenario with 100 families, 3 COs per family, 1000 projects, a single feasible start time for each project (i.e., $P = 1$), and a planning period of 25 years. In this case, the number of variables is $(100 \times 3) + (1000 \times 25) = 25300$ and the number of constraints is $100 + (1000 \times 1) + (100 \times 3) + 25 = 1425$.

3.2 NP-Hardness

It can be easily shown that this model is NP-hard by reduction from the MCKP [23] as follows. Given an instance of the MCKP with n classes and capacity C, a single-period instance of the proposed model can be constructed by assuming that the n classes refer to distinct families with each item in the class constituting a CO, and the budget being taken as the capacity C. Without loss of generality, assuming that all COs correspond to investment options, such that their weights and values are positive, then this instance is equivalent to the MCKP, which is known to be NP-hard. Therefore, the proposed problem is also NP-hard.

3.3 Sample Problem Data

An example of project data can be found in Table 1 while example CO data can be found in Table 2. These examples were arbitrary designed and do not reflect real projects or options. Furthermore, it is assumed that these projects have fixed start times, hence, the value of each CO is fixed. In Table 1, projects P1 through P5 represent investment options, which consume existing funds if selected, while projects P6 through P10 represent divestment options, which release funds if selected. In Table 2, COs are grouped into families, such that one CO is to be selected from each family. Note that, families F1 through F3 represent COs that correspond to investments, while F4 and F5 represent divestment options. In all cases, option $CO_{x,0}$ represents the "empty" option, which corresponds to no new project investments or divestments.

Table 1 Sample project data

Cost by year					
Item	1	2	3	4	5
P1	1	1	3	2	1
P2	0	8	12	9	1
P3	1	3	5	8	4
P4	0	0	9	12	5
P5	4	8	3	1	0
P6	−1	−12	−3	0	0
P7	−8	−6	−2	0	0
P8	−3	0	−1	−1	0
P9	0	0	−9	−2	−1
P10	−1	−3	−5	−2	−1

Table 2 Sample family and capability option data

Family	Option	Value	Projects
F1	$CO_{1,0}$	0	{}
	$CO_{1,1}$	0.3	{P1, P2}
	$CO_{1,2}$	0.5	{P1, P3}
	$CO_{1,3}$	0.6	{P1, P2, P3}
F2	$CO_{2,0}$	0	{}
	$CO_{2,1}$	0.2	{P4}
	$CO_{2,2}$	0.4	{P2, P4}
	$CO_{2,3}$	0.5	{P1, P2, P4}
	$CO_{2,4}$	0.6	{P1, P2, P5}
F3	$CO_{3,0}$	0	{}
	$CO_{3,1}$	0.7	{P1, P3, P4}
F4	$CO_{4,0}$	0	{}
	$CO_{4,1}$	-0.3	{P6}
	$CO_{4,2}$	-0.5	{P6, P8}
	$CO_{4,3}$	-0.8	{P6, P8, P10}
F5	$CO_{5,0}$	0	{}
	$CO_{5,1}$	-0.3	{P6, P7}
	$CO_{5,2}$	-0.6	{P7, P8, P9}
	$CO_{5,3}$	-0.7	{P6, P9, P10}

3.4 Similarity to Existing Problems

It is evident from the model formulation in Sect. 3 that the proposed problem bears similarity to multiple variants of the knapsack problem (see Sect. 2). The yearly cost vectors associated with each project resembles the MKP. The requirement for the selection of exactly one CO from each family is similar to the formulation of the MCKP. The combination of the yearly cost vectors and selection requirements shares commonalities with the MMKP. Finally, the COs being groupings of projects, where the total cost is dictated by the union of the selected projects, bears resemblance to the SUKP. Thus, the proposed problem can be viewed, in a simplistic manner, as a combination of these three knapsack variants—a problem which has not, to the knowledge of the authors, been considered in the literature. However, it should also be explicitly noted that the proposed problem has notable differences that prevent a knapsack formulation from adequately modeling this problem. As an example, there is no capacity *per se* in the proposed problem formulation, rather the budget available for funding investments is determined by the selection of divestment options, plus an (optional) allowable yearly budget deviation. Moreover, the flexibility in project start times means that there is also an associated scheduling component, further increasing the difficulty associated with optimizing the proposed model.

4 Heuristic Solution Approach

While the general structure of this problem resembles that of the MMKP and SUKP, the problem is sufficiently different such that existing heuristics for these problems cannot be directly used. Specifically, the lack of a fixed budget (i.e., capacity) and the existence of divestments (i.e., negative costs/values) are features that are explicitly excluded from existing heuristic techniques. Thus, there is a need to develop a customized heuristic for this problem. Nonetheless, significant inspiration was drawn from existing literature, notably the techniques employed by [21, 32].

The heuristic for the problem considered in this study is developed as follows. First, there must be a selection of divestment COs, such that an initial "budget" is generated. Next, there must be a selection of investment COs, which must respect the amount of funds made available through divestments and the allowable budget deviation. Finally, there should be a refinement process to improve the candidate solution, if necessary. Note that, there are a number of possibilities for each step and special care must be taken to ensure that the yearly spending constraints are not violated and that project costs are not counted multiple times.

In more detail, the proposed heuristic technique to generate a feasible solution can be summarized as:

1. Determine the initial divestment COs using a divestment selection strategy:
   - Select the CO from each (divestment) family that releases the most funds.
   - Select the CO from each (divestment) family with the largest ratio of funds released to value lost.
   - Randomly select a CO from each (divestment) family.
2. For each (investment) family where no selection has been made:
   a. Calculate the value-to-cost ratio for all COs. Note, this must account for projects that may have already been selected as part of other COs.
   b. Select the CO with the highest value-to-cost ratio that also leads to a feasible solution. Alternatively, select the project with the highest value, irrespective of its cost, that leads to a feasible solution. In either case, a project can only be selected if its inclusion in the portfolio does not lead to a violation of the budgetary constraint, including any allowable deviation.
3. Repeat Step 2 until a CO has been selected from each family.
4. Perform a swapping local search for refinement, as defined by
   a. For each family, determine the (feasible) CO swap that will lead to the largest increase in value.
   b. Perform the swap that leads to the largest increase over all families.
5. Repeat Step 4 until no improvements can be made or, alternatively, for a set number of iterations.

This heuristic process is shown in Algorithm 1 with the local search process shown in Algorithm 2. Note that, due to the requirement in Step 2b that the selection of a CO can only violate the (soft) budget constraint within the allowable deviation (BD_k), this procedure will always result in a feasible solution being returned. Recall that there is always a CO that has no cost or value in each family, thereby ensuring that even after the entire budget has been consumed, the "empty" CO can still be selected for the remaining families. In the context of this study, Step 4 is repeated until no improvements can be made.

The heuristic process in Algorithm 1 is then run multiple times and returns the portfolio with the best fitness. Specifically, the heuristic is repeated (once) for each combination of divestment and investment selection strategies, with the exception of the random divestment selection strategy, which is run some fixed number of times. In this study, the random divestment selection strategy is repeated 100 times, regardless of the number of divestment families. Note that, the heuristic process is deterministic unless the divestment selection strategy is random, hence the random strategy is run multiple times.

Algorithm 1 Heuristic Approach

```
function HEURISTIC(DS, IS)    ▷ DS: divestment selection strategy, IS: investment selection strategy
    p ← PORTFOLIO( )
    p ← SELECTDIVESTMENTS(DS)
    p ← SELECTINVESTMENTS(IS)
    p ← LOCALSEARCH(p)
    return p
end function
```

Algorithm 2 Local Search

```
function LOCALSEARCH(p)                                   ▷ p: portfolio
    repeat
        for all f ∈ F do                                  ▷ For each family
            p* ← p
            j ← LOCALSWAPSEARCH(f)                        ▷ find best item to swap in family f
            p_f ← SWAP(p, f, j)                           ▷ replace selection from family f with item j
            if f(p_f) > f(p*) then
                p* ← p_f
            end if
        end for
        p ← p*                                            ▷ perform the best swap over all families
    until STOPPINGCONDITION( )                            ▷ in this study, until no improvement is made
    return p
end function
```

5 Experimental Design

This section describes the experimental design used in this study. It should be explicitly noted that, while the problem instances do incorporate variable values for the COs, based on the various feasible start times of each project, the optimization process did not consider shifting of projects. In other words, it was assumed that for all projects p, $EST_p = LST_p$ and thus the value of each CO was fixed.

5.1 Synthetic Problem Instance Generation

This section describes the process by which the synthetic problem instances were generated. The data generation process was developed to provide similar overall characteristics to the real-world data.

5.1.1 Project Generation

The cost for each project follows a normal distribution with a random parameterization, hence the parameters must first be randomly chosen. First, define the current year as y,[1] the mean year (μ_y) as a random integer between ($y - 10$) and ($y + 20$), the standard deviation parameter (sd_y) as a random integer between 1 and 10, the scale parameter (s) as a random integer between 1000 and 15000, and the randomness parameter (r) as a random integer between 100 and the value drawn for s. The project start time is taken as

$$\max(\mu_y - 1.5sd_y, y),$$

while the end year is dictated by the project expenditure profile. The number of years until the project delivers capability (i.e., its effective time te_p) is sampled as a random integer between 3 and the number of years in the considered budget period. In the context of this study, a budget period of 25 years is considered.

To generate the cost, first define $m = s/r$ as the ratio between the the scale and the randomness parameter. For each year between y and the start year, the cost is defined as zero. Then, for each year t between the start year and the mean year μ_y, the cost is generated as

$$CP_{pt} = \max\left(\text{PDF}\left(\mathcal{N}(\mu_y, sd_y), i\right) \times s + U(-m...m), CP_{p(t-1)}, 0\right), \quad (14)$$

where $\mathcal{N}(\mu_y, sd_y)$ is the normal distribution with mean μ_y and standard deviation sd_y and PDF is the probability density function. For each year t (in descending order) between the number of budget periods and the mean year, the cost is generated as

[1] For the purposes of this study, the current year is taken as 2021.

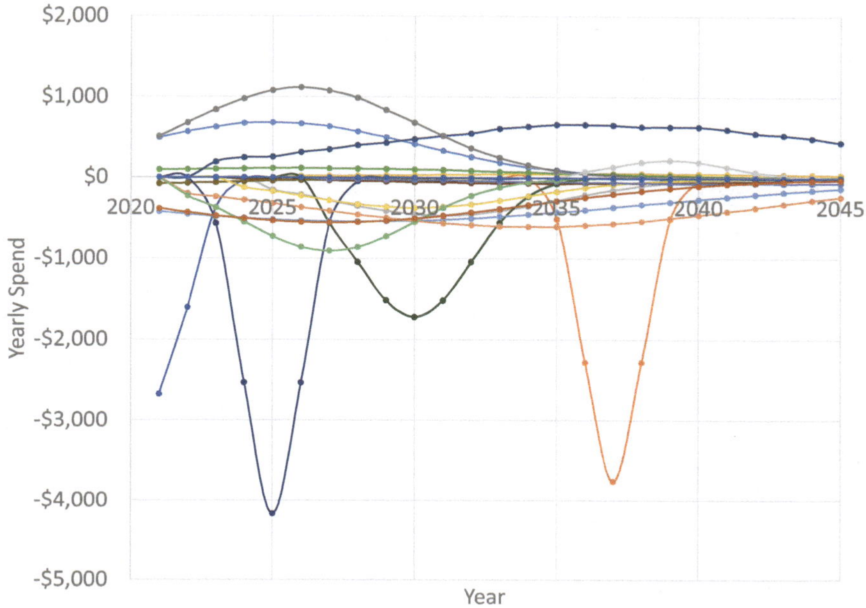

Fig. 1 20 randomly-generated project spend profiles

$$CP_{pt} = \max\left(\text{PDF}\left(\mathcal{N}(\mu_y, sd_y), i\right) \times s + U(-m...m), CP_{p(t+1)}, 0\right). \tag{15}$$

Note that, the difference between Eqs. (14) and (15) is that in Eq. (14), the cost is bounded by the previous year while in Eq. (15), the cost is bounded by the next year. Hence, this provides an increasing cost model until the mean year, after which the cost decreases each year until the project is completed. Finally, with 50% probability, the cost is negated to form a divestment project. An example of the project spend profiles for 20 projects is given in Fig. 1.

5.1.2 Family and Capability Option Generation

Assuming that a sufficient number of projects have already been generated, the following process is used to group the projects into families and COs. Firstly, the projects are split into two groups, namely divestments (i.e., net cost < 0) and investments (i.e., net cost > 0). A parameter, p_d, specifies the proportion of families that correspond to divestments. For each family, the number of COs is randomly chosen between 2 and 6. Then, a group of g projects, either all investment or all divestment, is selected. In the context of this study, g was taken to be 6. For each CO in the family, except the first, which is always taken as having no projects (i.e., the "empty" CO), a randomly-selected, unique, non-empty subset of the g projects is chosen.

Table 3 Parameter values for the problem generation process

Parameter	Value
Families	{25, 50, 100}
Projects	{100, 250, 500, 1000}
Divestments	{25%, 33%, 50%}
(Budget) Deviation	{\$0, \$500, \$1,000}

The value for each CO is determined by epoch in which the latest effective time associated with each of the constituent projects belongs. However, due to the variable start times of projects, it is possible that the value of a CO can change based on the selected start times for each project. In the context of this study, an epoch is taken as a 4 year period. To generate the value associated with each non-empty COs, an upper bound is first selected between 1.5 and 4. Then, for each epoch in the planning period, a value is randomly generated between 0 and the aforementioned upper bound. For divestment COs, values are negated to represent the loss of capabilities. Values are then sorted in descending order for investment COs, such that earlier investment is preferred, and ascending order for divestment COs, such that later divestment is preferred.

5.2 Problem Instances

A total of 108 problem instances were generated as all possible combinations of the parameters in Table 3. Scaling the number of families provides a direct scaling of the problem size. As the number of projects increases, the probability of overlap among different capability options decreases due to the grouping process described in the previous section. Hence, the overall problem difficulty should, in theory, decrease as the number of projects increases. The divestments parameter represents the proportion of families that contain only divestment options. Finally, the budget deviation is the amount by which the budget constraint can be violated in each year (i.e., BD_k from Eq. (10)). Thus, higher values for deviation provide a greater relaxation of the budget constraint, thereby decreasing the difficulty of the problem.

5.3 Algorithmic Control Parameters

The GA used a discrete representation, whereby a value of i at index f represents that capability option with index i was selected for family f. For example, the vector [1, 3, 0, 3] indicates that CO 1 was selected for family 1 (i.e., $O_{1,1}$), CO 3 was selected for family 2 (i.e., $O_{2,3}$), CO 0 (the "empty" option) was selected for family 3 (i.e., $O_{3,0}$),

Table 4 Control parameter values for the meta-heuristic approaches

Approach	Parameter	Value
GA	Population size	100
	Crossover	Half-Uniform (HUX)
	Crossover rate	0.9
	Mutation	(Integer) Polynomial
	Mutation rate	0.1
BRKGA	Population size	100
	Bias	0.7
	Proportion of elites (p_e)	0.25
	Proportion of mutants (p_m)	0.1
DE	Population size	100
	Differential weight (F)	0.5
	Crossover rate (CR)	0.3

and CO 3 was selected for family 4 (i.e., $O_{4,3}$). The BRKGA and DE approaches, which are inherently real-valued, used an alternative representation whereby the values at each index were normalized between 0 and 1. To decode the real-valued representation into the discrete representation used by the GA, each value was taken as the largest integer less than (i.e., the floor of) the solution vector multiplied by the number of COs for the corresponding family. For example, assuming that the number of COs for each family (NP) are [2, 3, 5, 6] and the solution vector is [0.60, 0.50, 0.05, 0.65], the decoded solution would be the same as the previous example, i.e., [1, 3, 0, 3].

Each of the meta-heuristic algorithms were seeded with either the best heuristic solution provided by Algorithm 1 if a positive fitness value was attained, or the zero vector, which represents the base-case feasible solution of all "empty" COs. A preliminary study found that this seeding strategy improved GA solutions by an average of 10.2% [18]. For the real-valued algorithms (BRKGA and DE), the seed solution was first normalized to values between 0 and 1 by dividing the solution vector by the total number of capability options for each family, as the reverse of the decoding process above. The meta-heuristics were run until 100 iterations had occurred with no improvement to the global best fitness. Performance data is collected over 3 0 independent runs for each approach.

Table 4 provides the algorithmic control parameter values employed by each of the meta-heuristic approaches. All control parameter values were taken as the default values from the *pymoo* package [2]. For DE, the rand/1/bin strategy was employed along with an adaptive differential weight strategy, whereby the differential weight was randomly modified using a uniform distribution in the range of [0, 1] applied independently to each individual [3]. For BRKGA, the proportion of offspring was taken as $1 - p_e - p_m$, where p_e and p_m are the proportion of elites and mutants, respectively.

5.4 Statistical Analysis

The metric used in the statistical analysis process is the rank corresponding to the average portfolio value across each problem instance across all independent runs. Ranks are determined in ascending order, such that a rank of 1 corresponds to the highest portfolio value for a particular problem instance. To identify the pairwise differences among the examined approaches, Friedman's test for multiple comparisons among all methods [9, 10] with Shaffer correction [28] was employed. The significance level was set at $\alpha = 0.01$.

To visualize the statistical results, critical difference plots [5] were generated for each statistical comparison. The critical difference denotes the difference in average rank across a set of problems that must be observed for the performance of two approaches to be considered significantly different among the considered problems. A critical difference plot is produced by placing the considered approaches on a horizontal axis such that their positions correspond to their respective average rankings. Two (or more) approaches are grouped by a line if the difference between their average ranks was less than the calculated critical difference. That is, approaches appearing to the left demonstrated superior performance to those on the right, on average, while grouping by a line denotes that no statistically significant difference in performance was observed among the approaches within that group. It should be noted that, in general, the critical difference relationship is not transitive.

6 Results

This section presents the experimental results and provides a discussion of their significance.

6.1 Validating the Solution Approaches

To first validate the solution approaches, an exhaustive solver was implemented to determine exact solutions. However, due to the time complexity associated with exhaustively searching all possible solutions, only small instances (i.e., fewer than 10 families) can be reasonably addressed. Table 5 provides the results comparing the performance of each approach with that of the exhaustive solver. Results are expressed as a percentage of the exact fitness attained, averaged across 30 problem instances. In the case of the meta-heuristic approaches, results for each problem are taken as the average across 30 independent runs.

It is evident from Table 5 that the heuristic approach is capable of finding reasonable portfolios in most scenarios—in 25 of 36 problem configurations, the heuristic approach attained an average of at least 70% of the fitness associated with the exact

Table 5 Accuracy of heuristic and meta-heuristic approaches. Results are expressed as a percentage of the exact fitness, averaged across 30 problem instances

				Heuristic		GA		BRKGA		DE	
Families	Projects	Divestment (%)	Deviation	Mean (%)	Std. (%)	Mean (%)	Std. (%)	Mean (%)	Std. (%)	Mean (%)	Std. (%)
8	50	25	0	65.3	44.7	99.6	1.5	100.0	0.0	99.8	0.7
	50	25	500	60.7	42.4	99.7	0.8	99.7	1.3	99.7	0.7
	50	25	1000	62.5	38.3	99.8	0.7	100.0	0.1	99.4	1.7
	50	33	0	72.2	38.1	99.8	0.8	100.0	0.0	99.5	1.6
	50	33	500	81.7	30.1	99.8	1.0	100.0	0.0	99.2	3.2
	50	33	1000	71.8	36.3	99.4	2.3	99.9	0.3	98.7	3.1
	50	50	0	75.3	39.9	100.0	0.0	100.0	0.1	99.6	0.9
	50	50	500	63.1	42.2	99.7	1.5	100.0	0.0	99.0	4.9
	50	50	1000	70.0	40.4	100.0	0.1	100.0	0.0	99.9	0.5
	100	25	0	89.9	22.9	99.7	1.5	100.0	0.0	99.7	1.2
	100	25	500	65.4	41.2	99.3	3.0	100.0	0.1	99.5	1.4
	100	25	1000	72.7	35.9	99.3	2.5	100.0	0.2	99.1	2.2
	100	33	0	89.9	26.3	100.0	0.0	100.0	0.0	99.8	0.8
	100	33	500	69.1	38.9	99.9	0.5	100.0	0.0	100.0	0.1
	100	33	1000	84.0	26.4	99.3	2.6	99.9	0.3	99.2	2.4
	100	50	0	75.3	39.5	99.9	0.3	100.0	0.0	99.8	0.8
	100	50	500	57.7	45.3	100.0	0.0	99.3	3.6	99.8	0.7
	100	50	1000	55.8	39.1	99.8	0.9	99.9	0.6	99.8	0.8
9	50	25	0	88.0	24.9	99.9	0.7	100.0	0.0	99.5	1.7
	50	25	500	70.8	34.9	97.2	8.0	99.3	2.0	97.2	7.5
	50	25	1000	83.0	28.6	99.7	0.9	99.9	0.4	99.9	0.5
	50	33	0	76.1	39.4	98.8	5.6	98.7	5.4	98.3	5.9
	50	33	500	73.8	33.9	98.7	3.5	99.5	2.1	98.9	3.6
	50	33	1000	75.3	28.5	99.0	3.2	100.0	0.2	99.4	1.2
	50	50	0	74.7	41.5	100.0	0.0	99.3	3.6	99.1	3.4
	50	50	500	68.3	37.2	99.0	4.0	99.7	1.3	98.0	5.4
	50	50	1000	57.5	42.2	99.9	0.5	100.0	0.1	99.7	0.7
	100	25	0	77.4	39.6	99.7	1.6	100.0	0.0	99.6	1.7
	100	25	500	73.9	36.0	99.5	2.3	99.5	2.1	97.9	5.6
	100	25	1000	75.0	29.5	99.5	1.4	99.9	0.3	99.4	1.6
	100	33	0	84.0	29.1	99.5	2.0	100.0	0.0	99.7	1.1
	100	33	500	76.7	32.8	99.6	1.0	99.9	0.4	99.1	1.8
	100	33	1000	76.6	32.8	98.3	5.9	99.9	0.2	99.2	3.6
	100	50	0	66.9	43.2	99.8	0.6	99.9	0.3	99.7	0.8
	100	50	500	72.7	27.9	99.7	1.3	99.8	0.6	99.7	0.7
	100	50	1000	70.2	38.1	99.8	0.9	100.0	0.0	99.8	0.6

solution. The minimum average attained by the heuristic approach was 55.8% of the exact fitness, whereas the maximum was 89.9%. While not shown in Table 5, the heuristic was able to achieve the optimal fitness in at least one problem instance for nearly all configurations. Conversely, there was typically at least one instance where the heuristic approach failed to find a feasible solution. Given that the failure to find a feasible solution occurred on very few instances for a given parameter configuration, it is hypothesized that this can be attributed to anomalous behavior in either the heuristic or the data generation process rather than as a result of the problem characteristics. As expected, there is a general tendency for the heuristic solutions to degrade in quality as the number of projects decreased (indicating a higher degree of overlap among different COs) or the proportion of divestment families was increased. From the results in Table 5, it is clear that the performance of the heuristic was not strikingly different when increasing the number of families from 8 to 9.

Considering the meta-heuristic approaches, Table 5 depicts that across all problem configurations, a minimum fitness of 97.2% was reached by the GA and DE approaches, whereas the BRKGA approach achieved a minimum of 98.7%. These results indicate that, for every problem configuration examined, nearly all of the meta-heuristic runs resulted in the optimal fitness. However, it is important to note that a result less than 100% indicates that there was at least one run that resulted in a sub-optimal fitness. Overall, these results highlight the efficacy of the meta-heuristic approaches with regards to improving the heuristic solutions.

6.2 Effect of Seeding

To ascertain whether the heuristic seeding process added a significant performance improvement to the meta-heuristic approaches, experiments were also conducted with no seeding. That is, the meta-heuristics were also run with a purely random initialization procedure. Table 6 presents a summary of the average improvement for each meta-heuristic as a result of including a seeded solution to the initial population. From these results, it was found that the GA was improved by 10.27%, BRKGA was improved by 1.04%, and DE was improved by 3.68% when averaged across all 108 problem instances. Notably, the margin of improvement for the seeded variants of

Table 6 Average improvement for each meta-heuristic by including a single seeded solution in the initial population

Families	GA (%)	BRKGA (%)	DE (%)
25	4.98	3.18	1.23
50	12.78	2.56	4.12
100	13.04	−2.63	5.70
Overall	10.27	1.04	3.68

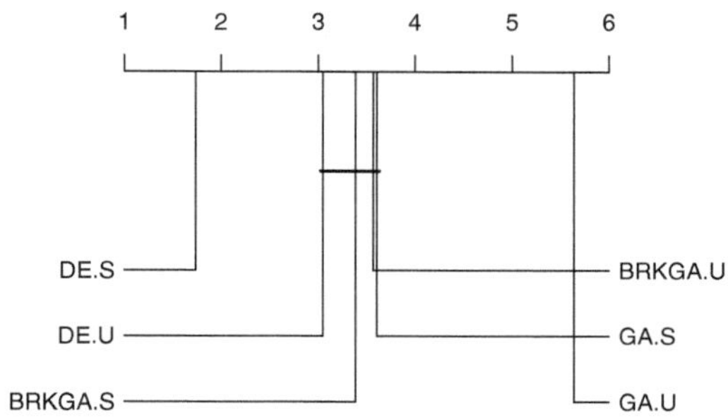

Fig. 2 Critical difference plot comparing seeded (S) and unseeded (U) meta-heuristics over all 108 problem instances

GA and DE increased when larger problem instances were considered. This provides evidence that the benefits of seeding were increased for problems with greater difficulty. In contrast, the effect of seeding on BRKGA decreased when considering larger problem instances.

Figure 2 presents the critical difference plot comparing the seeded (S) and unseeded (U) variants across all 108 problem instances. For each of the three meta-heuristics, it was found that the seeded variant demonstrated better performance. For both the GA and DE approaches, the effect of seeding was found to be statistically significant. In contrast, despite the better performance exhibited by BRKGA with seeding, the effect was not found to be statistically significant.

6.3 Main Results

The results in Sect. 6.1 provided a preliminary comparison of the heuristic and meta-heuristic approaches to an exhaustive solver on small problem instances. In this section, the performance of the proposed solution approaches is presented and analyzed in greater detail on larger problem instances. As previously noted, the scale of these problem instances is too large for the exhaustive solver to consider in a reasonable amount of time.

Tables 7, 8, and 9 provide the numerical results for problem instances with 25, 50, and 100 families, respectively. Results are expressed as the total portfolio value according to Eq. (12). It should be noted that each problem instance is completely independent and portfolio values can not be directly compared between problem instances. From these tables, a few immediate observations can be made. In terms of overall performance, the average fitness attained by DE was generally greater than that attained by the other approaches. This is reinforced by the critical different plot in Fig. 3 corresponding to a statistical significance test across all problem instances, which depicts the performance of DE as exhibiting a statistically significant difference when compared to the other approaches. Moreover, Fig. 3 depicts that no statistically significant difference was observed between the performance of GA and

Table 7 Total portfolio value for the heuristic and meta-heuristic approaches on problem instances with 25 families

				GA		BRKGA		DE	
Projects	Divestment (%)	Deviation	Heuristic	Mean	Std.	Mean	Std.	Mean	Std.
100	25	$0	0.62	0.65	0.07	0.69	0.08	0.68	0.07
	25	$500	5.56	5.56	0.00	5.56	0.00	5.56	0.00
	25	$1,000	1.45	3.39	0.68	3.64	0.09	3.62	0.07
	33	$0	8.13	8.73	0.38	8.99	0.01	8.82	0.30
	33	$500	7.91	8.31	0.25	8.32	0.30	8.32	0.20
	33	$1,000	13.11	13.11	0.00	13.11	0.00	13.11	0.00
	50	$0	3.91	4.94	0.32	4.88	0.33	5.08	0.18
	50	$500	9.12	9.12	0.00	9.12	0.00	9.12	0.00
	50	$1,000	0.91	2.83	0.36	3.02	0.12	3.09	0.11
250	25	$0	2.82	3.42	0.37	3.70	0.16	3.65	0.22
	25	$500	9.46	9.47	0.03	9.47	0.03	9.49	0.06
	25	$1,000	10.15	10.95	0.27	10.97	0.26	11.09	0.07
	33	$0	10.55	12.09	0.10	12.10	0.29	12.11	0.08
	33	$500	8.12	8.97	0.41	9.23	0.38	9.34	0.25
	33	$1,000	6.97	7.86	0.29	8.15	0.14	7.89	0.32
	50	$0	5.63	8.18	0.59	8.56	0.28	8.71	0.18
	50	$500	3.09	5.23	0.57	5.59	0.10	5.62	0.11
	50	$1,000	12.45	14.10	0.29	14.05	0.22	14.13	0.16
500	25	$0	5.00	5.73	0.10	5.74	0.08	5.71	0.22
	25	$500	2.94	4.75	0.69	5.22	0.30	5.06	0.42
	25	$1,000	8.17	8.46	0.06	8.47	0.10	8.48	0.05
	33	$0	11.29	11.31	0.03	11.30	0.03	11.31	0.04
	33	$500	2.55	4.70	0.25	4.70	0.39	4.79	0.07
	33	$1,000	9.08	10.37	0.72	10.66	0.80	11.19	0.45
	50	$0	5.36	6.68	0.24	6.71	0.16	6.75	0.12
	50	$500	6.40	6.67	0.21	6.65	0.22	6.79	0.20
	50	$1,000	1.44	4.78	0.55	4.97	0.37	5.12	0.24
1000	25	$0	6.57	6.57	0.00	6.57	0.00	6.57	0.00
	25	$500	4.39	5.29	0.14	5.49	0.21	5.43	0.20
	25	$1,000	8.89	9.57	0.16	9.56	0.17	9.49	0.21
	33	$0	8.07	9.36	0.35	9.43	0.37	9.67	0.32
	33	$500	9.77	11.40	0.22	11.58	0.00	11.49	0.17
	33	$1,000	4.02	5.14	0.29	5.16	0.21	5.16	0.21
	50	$0	3.36	4.03	0.20	4.10	0.16	4.06	0.14
	50	$500	6.70	7.35	0.25	7.44	0.12	7.41	0.19
	50	$1,000	6.18	6.44	0.13	6.47	0.10	6.48	0.08

BRKGA across all problem instances, despite the average rank for GA being slightly lower.

In nearly all cases reported in Tables 7, 8, and 9, the meta-heuristic approaches were able to effectively improve the initial heuristic solutions. When averaged across

Table 8 Total portfolio value for the heuristic and meta-heuristic approaches on problem instances with 50 families

				GA		BRKGA		DE	
Projects	Divestment (%)	Deviation	Heuristic	Mean	Std.	Mean	Std.	Mean	Std.
100	25	$0	15.62	15.66	0.05	15.62	0.02	15.67	0.05
	25	$500	18.67	18.73	0.18	18.69	0.13	18.82	0.24
	25	$1,000	22.35	22.39	0.06	22.35	0.00	22.37	0.09
	33	$0	13.43	13.92	0.32	13.65	0.34	14.45	0.32
	33	$500	15.60	16.26	0.75	16.90	0.81	17.11	0.44
	33	$1,000	6.33	8.50	0.66	9.34	0.98	9.66	0.57
	50	$0	7.10	9.23	0.53	9.98	0.68	10.78	0.41
	50	$500	0.00	3.09	0.30	3.26	0.16	3.20	0.17
	50	$1,000	16.15	18.21	0.58	18.48	0.90	18.85	0.75
250	25	$0	14.77	16.03	0.40	15.23	0.70	16.08	0.44
	25	$500	14.85	15.06	0.47	15.11	0.47	16.18	0.74
	25	$1,000	14.41	16.02	0.45	15.97	0.93	16.64	0.53
	33	$0	14.19	15.80	0.67	16.14	0.64	16.50	0.28
	33	$500	18.16	19.21	0.72	19.26	0.75	19.89	0.22
	33	$1,000	17.91	18.65	0.19	18.24	0.39	18.59	0.37
	50	$0	2.55	5.27	0.78	5.29	0.91	6.05	0.41
	50	$500	7.73	14.64	0.86	15.11	0.74	15.98	0.37
	50	$1,000	10.38	13.06	0.54	13.44	0.40	13.56	0.30
500	25	$0	14.12	14.35	0.35	14.21	0.28	14.66	0.43
	25	$500	18.81	18.91	0.22	18.81	0.01	19.10	0.44
	25	$1,000	13.30	15.11	0.63	15.57	0.65	15.97	0.40
	33	$0	11.99	13.95	0.44	14.16	0.67	14.33	0.65
	33	$500	20.95	21.36	0.27	21.17	0.29	21.75	0.29
	33	$1,000	13.90	16.18	1.05	17.30	0.64	17.55	0.54
	50	$0	3.96	10.79	0.69	11.39	0.48	11.73	0.28
	50	$500	13.51	18.21	1.29	19.49	0.51	19.86	0.25
	50	$1,000	12.50	15.51	0.60	15.53	0.92	15.75	0.37
1000	25	$0	16.90	17.15	0.26	17.03	0.29	17.49	0.33
	25	$500	20.84	22.70	0.31	22.52	0.97	22.81	0.59
	25	$1,000	17.01	19.98	0.72	20.63	0.59	20.62	0.50
	33	$0	11.75	14.16	0.32	13.93	0.56	14.64	0.36
	33	$500	15.73	17.09	0.45	17.21	0.60	17.36	0.25
	33	$1,000	20.12	20.56	0.00	20.13	0.08	20.48	0.26
	50	$0	9.59	15.73	0.95	16.97	0.63	17.20	0.38
	50	$500	10.20	15.01	0.45	15.12	0.43	15.54	0.28
	50	$1,000	5.74	9.88	0.99	11.18	0.39	11.24	0.38

Table 9 Total portfolio value for the heuristic and meta-heuristic approaches on problem instances with 100 families

				GA		BRKGA		DE	
Projects	Divestment (%)	Deviation	Heuristic	Mean	Std.	Mean	Std.	Mean	Std.
100	25	$0	45.99	46.77	1.06	45.99	0.00	45.99	0.00
	25	$500	65.75	65.75	0.00	65.75	0.00	65.75	0.00
	25	$1,000	50.62	50.65	0.18	50.62	0.00	50.62	0.00
	33	$0	44.44	44.58	0.41	44.44	0.00	44.44	0.00
	33	$500	45.90	46.07	0.32	45.90	0.00	45.91	0.07
	33	$1,000	45.83	45.83	0.00	45.83	0.00	45.83	0.00
	50	$0	26.38	26.83	0.61	26.38	0.00	26.66	0.85
	50	$500	31.92	31.98	0.15	31.92	0.00	31.92	0.00
	50	$1,000	23.26	23.66	0.68	23.26	0.00	23.26	0.00
250	25	$0	49.19	49.39	0.42	49.19	0.00	49.25	0.23
	25	$500	53.49	55.19	0.76	53.49	0.00	55.32	1.18
	25	$1,000	66.07	67.09	0.49	66.07	0.00	66.92	1.25
	33	$0	26.31	27.87	0.93	26.31	0.00	27.66	1.95
	33	$500	37.85	39.21	1.22	37.85	0.00	42.86	1.83
	33	$1,000	43.58	44.41	0.85	43.58	0.00	43.58	0.00
	50	$0	17.85	19.77	0.53	18.66	1.65	23.40	0.57
	50	$500	34.23	37.26	0.60	34.60	0.97	38.52	1.33
	50	$1,000	17.86	22.53	2.28	23.87	4.02	28.58	0.74
500	25	$0	27.91	28.90	0.65	27.91	0.00	30.09	1.44
	25	$500	33.54	34.90	0.82	33.54	0.00	36.35	1.68
	25	$1,000	42.46	43.93	0.79	42.46	0.00	44.78	0.98
	33	$0	27.54	30.46	1.31	27.54	0.00	32.72	2.53
	33	$500	13.69	18.22	1.36	14.37	2.14	20.89	1.53
	33	$1,000	20.03	24.30	1.12	20.03	0.00	27.11	1.71
	50	$0	15.36	27.18	2.33	29.73	1.46	31.78	0.79
	50	$500	24.78	29.45	1.19	29.44	2.87	32.60	0.51
	50	$1,000	16.90	24.23	1.46	26.16	1.94	27.89	0.52
1000	25	$0	24.65	28.22	0.69	24.65	0.00	29.09	0.42
	25	$500	33.59	35.40	1.13	33.59	0.00	36.65	1.43
	25	$1,000	35.17	36.36	1.03	35.17	0.00	40.39	1.70
	33	$0	44.29	44.80	0.53	44.29	0.00	45.49	1.12
	33	$500	33.16	36.12	1.05	33.16	0.00	37.85	1.68
	33	$1,000	39.23	41.08	0.97	39.23	0.00	43.19	1.30
	50	$0	25.43	30.63	1.31	27.05	2.97	34.24	0.70
	50	$500	16.54	21.92	1.53	20.49	4.09	26.30	0.88
	50	$1,000	14.73	25.11	1.52	28.03	1.11	30.71	0.65

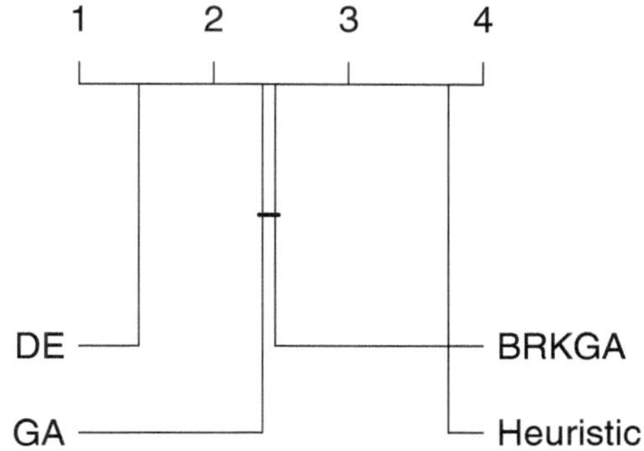

Fig. 3 Critical difference plot over all 108 problem instances

all problem instances, the fitness of the heuristic approach was 76.7% of the average fitness attained by GA, 71.1% of the average fitness attained by BRKGA, and 69.9% of the average fitness attained by DE. However, in most cases, the relative difference between the fitness attained by the heuristic approach and the meta-heuristics was rather small, while a few significant differences caused the overall average to decrease. Specifically, when compared to the DE results, the heuristic approach attained greater than 80% of the fitness attained by DE for 68.5% of problems, and greater than 90% of the fitness for 47.2% of problems. In other words, the best performing meta-heuristic approach led to improvements of less than 10% in 52.8% of problem instances and less than 20% in 31.5% of problems. The main observations from these results are that the heuristic is capable of generating good quality initial solutions and that the use of meta-heuristic approaches to improve the solution quality is warranted for this problem.

Another key observation is that the performance of the heuristic approach, relative to the meta-heuristics, is severely degraded when the proportion of families corresponding to divestments is 50%. However, this is not an unexpected result as the greedy nature of the heuristic approach leads to selection of a divestment option wherever possible. As a result, without a large proportion of investment options to select from, it is unlikely to find a high-valued set of investments within the available budget to offset the decrease in value.

In the following subsections, the effect of various problem characteristics is examined in greater detail.

6.3.1 Effect of the Number of Families

To ascertain the effect of the number of families on the relative performance, Figs. 4, 5, and 6 present the critical difference plot for the problem instances with 25, 50, and 100 families, respectively. Irrespective of the number of families, DE always led to the lowest mean rank (i.e., the best performance). When problem instances with 25 families were considered, there was no statistically significant difference detected

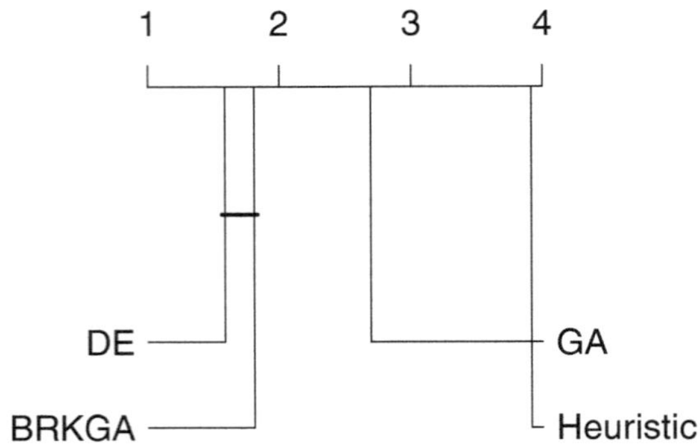

Fig. 4 Critical difference plot over all 36 problem instances with 25 families

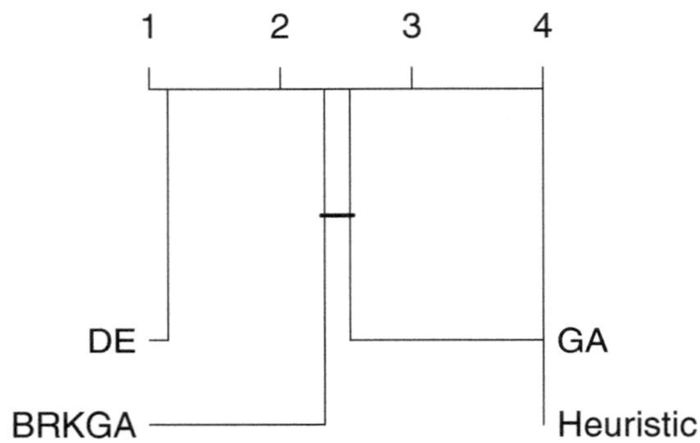

Fig. 5 Critical difference plot over all 36 problem instances with 50 families

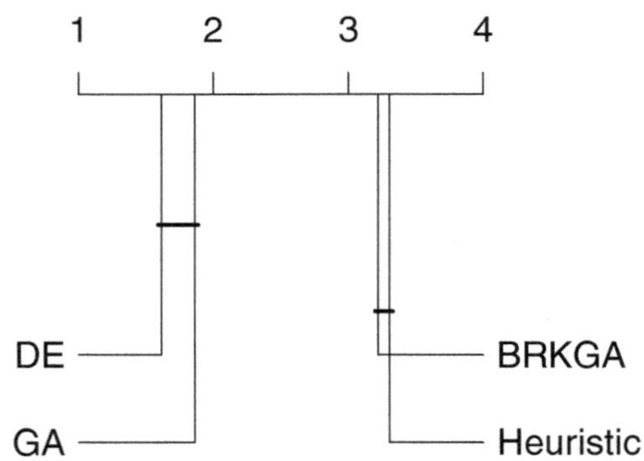

Fig. 6 Critical difference plot over all 36 problem instances with 100 families

between the performance of DE and BRKGA. When considering problem instances with 50 families, DE was superior to the other approaches, while there was no statistically significant difference detected between the performance of BRKGA and GA. For the problem instances with 100 families, there was no statistically significant difference detected between the performance of DE and GA while the BRKGA and heuristic approaches led to an insignificant difference in performance. This was

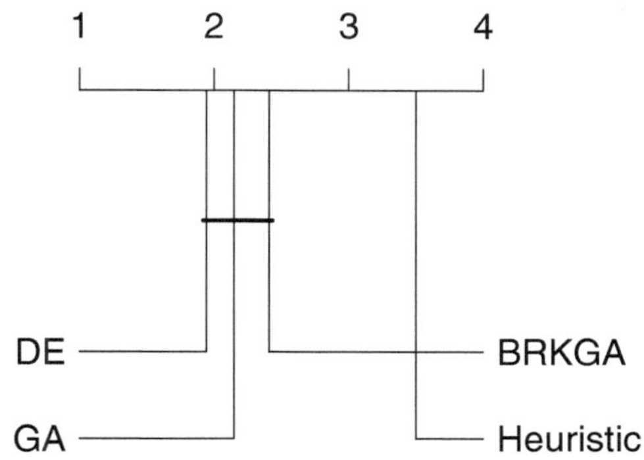

Fig. 7 Critical difference plot over all 27 problem instances with 100 projects

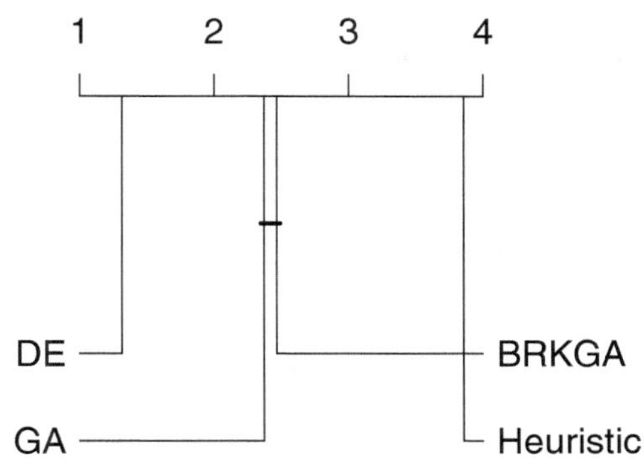

Fig. 8 Critical difference plot over all 27 problem instances with 250 projects

the only identified scenario in which the heuristic approach was not significantly outperformed by a meta-heuristic approach.

6.3.2 Effect of the Number of Projects

To ascertain the effect of the number of projects, Figs. 7, 8, 9 and 10 present the critical difference plot for the problem instances with 100, 250, 500, and 1000 projects, respectively. Again, in all cases, the DE approach led to the best performance, but its performance was insignificantly different from the performance of both GA and BRKGA on instances with 100 projects. For each set of problems with 250, 500, and 1000 projects, no significant difference was identified between the performance of the GA and BRKGA approaches.

6.3.3 Effect of the Proportion of Divestments

Figures 11, 12, and 13 present the critical difference plot for the problem instances with 25%, 33%, and 50% of families corresponding to divestment projects, respec-

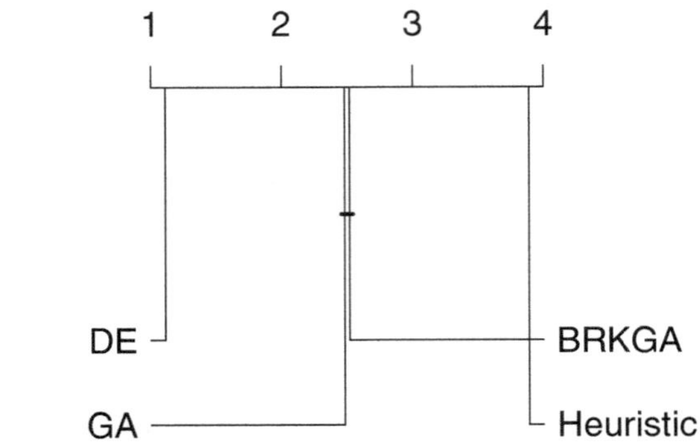

Fig. 9 Critical difference plot over all 27 problem instances with 500 projects

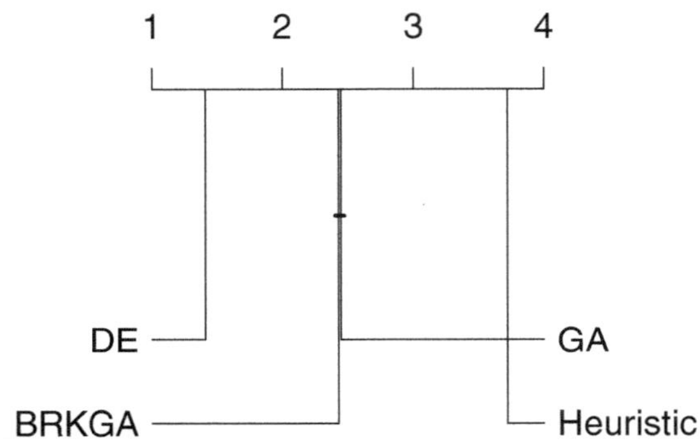

Fig. 10 Critical difference plot over all 27 problem instances with 1000 projects

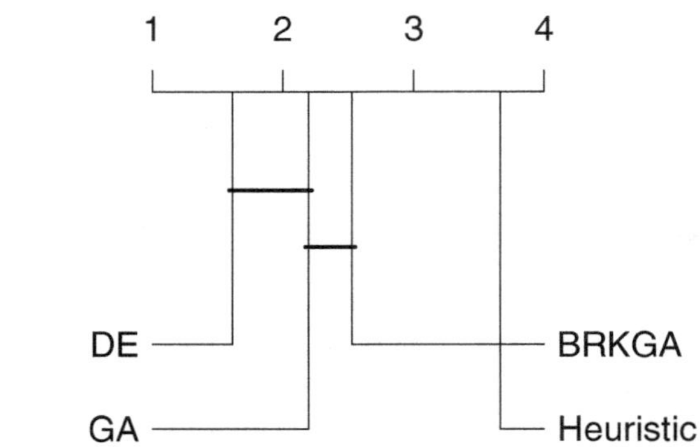

Fig. 11 Critical difference plot over all 36 problem instances with 25% divestment families

tively. In all cases, DE was again superior to the other approaches, but demonstrated no significant difference in performance when compared to GA on instances with 25% divestment-oriented families. In all cases, the GA and BRKGA approaches led to insignificantly different fitness values.

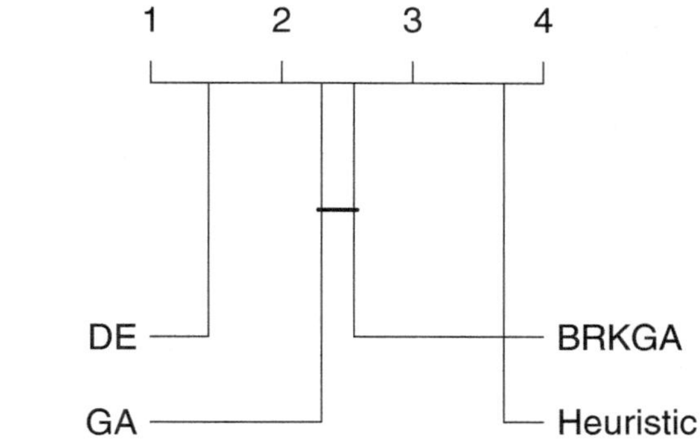

Fig. 12 Critical difference plot over all 36 problem instances with 33% divestment families

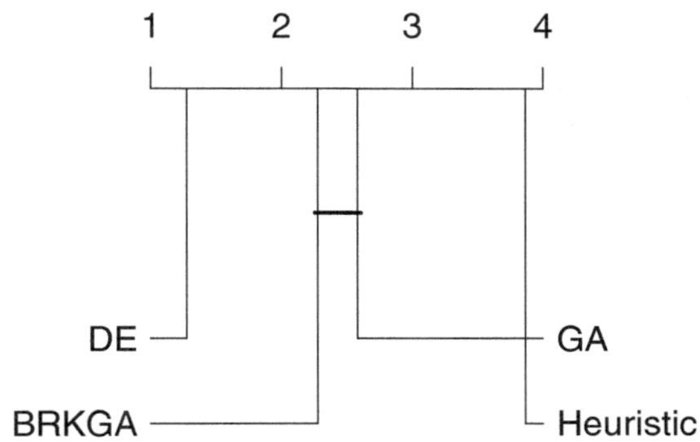

Fig. 13 Critical difference plot over all 36 problem instances with 50% divestment families

6.3.4 Effect of the Yearly Budget Deviation

To determine the effect of varying the allowable yearly budget deviation (i.e., BD_k), Figs. 14, 15, and 16 present the critical difference plot for the problem instances with a yearly budget deviation of \$0 (i.e., no allowable deviation), \$500, and \$1,000, respectively. In all cases, DE was superior to the other approaches and led to a statistically significant difference in performance, whereas the performance of the GA and BRKGA approaches was insignificantly different in all cases.

6.4 Summary

For all problem characteristics, the DE approach had the lowest mean rank and, in many cases, exhibited a statistically significant difference in performance when compared to the other meta-heuristics approaches; in only limited cases did the GA, BRKGA, or both exhibit insignificantly different performance compared to DE. In nearly all cases, the difference in performance between the GA and BRKGA approaches was not statistically significant. Thus, there is clear evidence to suggest

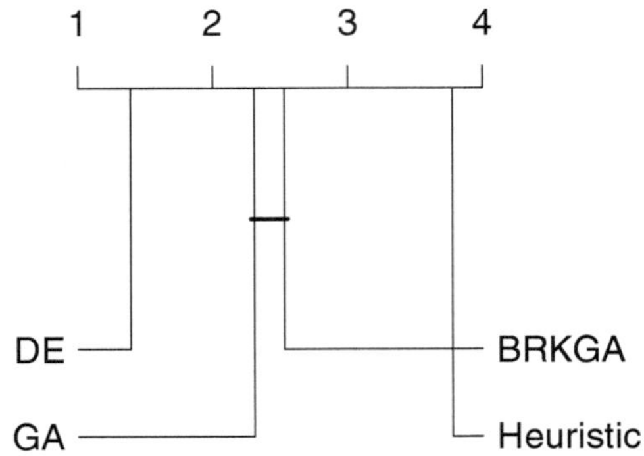

Fig. 14 Critical difference plot over all 36 problem instances with no allowable yearly budget deviation

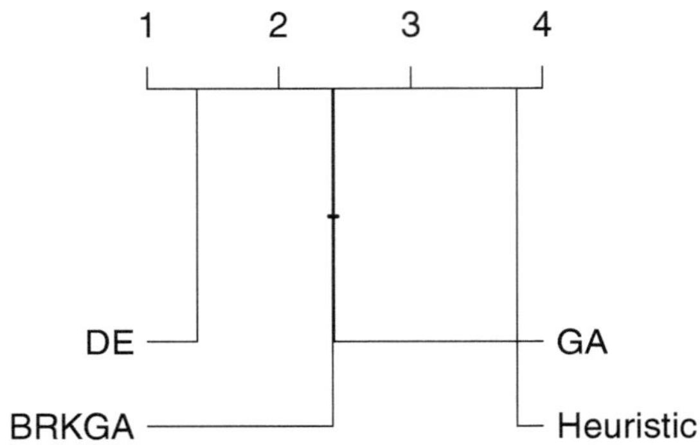

Fig. 15 Critical difference plot over all 36 problem instances with an allowable yearly budget deviation of $500

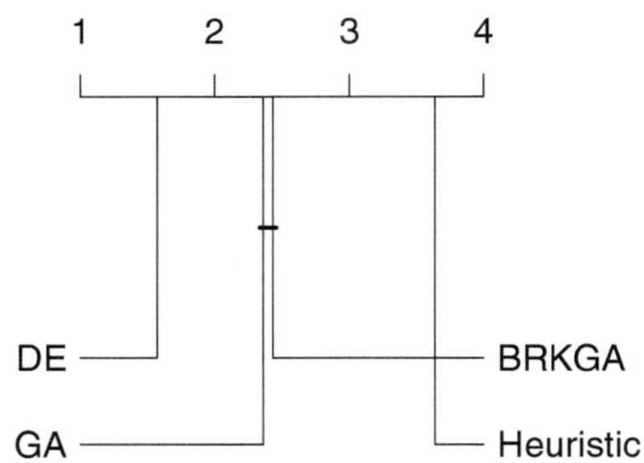

Fig. 16 Critical difference plot over all 36 problem instances with an allowable yearly budget deviation of $1,000

that the relative performance of these meta-heuristic approaches is not significantly influenced by the problem characteristics. However, it is again noted that no control parameter tuning was performed, nor was a search for the most effective operators performed. Therefore, it is certainly possible that the relative performance of the meta-heuristic approaches could be different if they were appropriately tuned, especially with regards to specific problem characteristics. It was observed that the performance of the heuristic approach was significantly worse than that of the meta-heuristic approaches in all cases except one. This is not unexpected given that the meta-heuristics were seeded with the heuristic solution and were predominately used to refine the initial solution. Nonetheless, the solutions returned by the heuristic approach were still of a reasonable quality and, in over 2 out of every 3 problem instances, the best meta-heuristic approach improved the heuristic solution by less than 20%.

7 Conclusions and Future Work

This study examined the selection and scheduling of a portfolio of projects in the context of future force design in the Australian Defence Force, whereby the projects are arranged into families, each consisting of subsets of projects, referred to as capability options. The overall objective is to maximize the portfolio value (i.e., the delivery of capabilities) by selecting a set of capability options while adhering to the operational constraints. Each project has an associated yearly spend profile, which may be either positive (for investment projects) or negative (for divestment projects), while the overall value of the portfolio is dictated by the implementation of various capability options. The selection process dictates that exactly one capability option must be selected from each family. That is, projects are not directly selected for inclusion in the portfolio, rather the set of projects implemented is defined by the union of the projects constituting each of the selected capability options. Furthermore, the cost associated with a project must only be counted once, regardless of the number of times it is selected for inclusion in the portfolio. Every family includes an "empty" capability option, which consists of the empty subset of projects and has no associated cost nor effect on the total portfolio value. Finally, the start time of individual projects can be adjusted, within a limited window, to provide earlier or later delivery of capabilities—this also shifts the project spend profile accordingly. However, the shifting of projects was not considered in this study.

While this problem bears some similarity to existing problems in the literature, notably the MMKP and SUKP, no existing problem adequately addresses all aspects of this problem. Therefore, this study formulated a mathematical model for the considered problem. A data generation process was developed to appropriately capture the characteristics of real-world data. A customized heuristic solver was designed to provide very rapid feasible solutions. These heuristic solutions were then improved using three meta-heuristic approaches. A set of 108 synthetic problem instances were generated to represent a wide variety of problem characteristics. Results indicated that the heuristic approach was able to provide high-quality solutions nearly instantaneously, while the meta-heuristic approaches could provide enhanced solutions within a reasonable amount of time.

First and foremost, future work will examine the shifting of project start times such that the value of each capability option will vary based on the scheduling of the constituent projects. Inclusion of dynamic start times will also warrant the development of special-purpose heuristics that address both the selection and scheduling aspects. Another critical avenue of future work lies in the appropriate tuning of the meta-heuristic approaches to further enhance their performance. Furthermore, the implementation of the proposed model using a commercial exact solver is needed as it would facilitate absolute performance comparisons on larger problem instances.

Acknowledgements This work was supported by the Australian Department of Defence, Defence Science and Technology Project RG191353.

References

1. Bean, J.C.: Genetic algorithms and random keys for sequencing and optimization. ORSA J. Comput. **6**(2), 154–160 (1994). https://doi.org/10.1287/ijoc.6.2.154
2. Blank, J., Deb, K.: Pymoo: Multi-objective optimization in python. IEEE Access **8**, 89497–89509 (2020). https://doi.org/10.1109/ACCESS.2020.2990567. https://ieeexplore.ieee.org/document/9078759/
3. Dawar, D., Ludwig, S.: Differential evolution with dither and annealed scale factor. In: 2014 IEEE Symposium on Differential Evolution, pp. 1–8. IEEE (2014). https://doi.org/10.1109/SDE.2014.7031528
4. De Spiegeleire, S.: Ten trends in capability planning for defence and security. RUSI J. **156**(5), 20–28 (2011). https://doi.org/10.1080/03071847.2011.626270
5. Demšar, J.: Statistical comparisons of classifiers over multiple data sets. J. Mach. Learn. Res. **7**, 1–30 (2006)
6. Department of Defence: 2016 Integrated Investment Program (2016)
7. Dixit, V., Tiwari, M.K.: Project portfolio selection and scheduling optimization based on risk measure: a conditional value at risk approach. Ann. Oper. Res. **285**(1–2), 9–33 (2020). https://doi.org/10.1007/s10479-019-03214-1
8. Fisher, B., Brimberg, J., Hurley, W.J.: An approximate dynamic programming heuristic to support non-strategic project selection for the royal Canadian navy. J. Def. Model. Simul. **12**(2), 83–90 (2015). https://doi.org/10.1177/1548512913509031
9. Friedman, M.: The use of ranks to avoid the assumption of normality implicit in the analysis of variance. J. Am. Stat. Assoc. **32**(200), 675 (1937). https://doi.org/10.2307/2279372. https://www.jstor.org/stable/2279372?origin=crossref
10. Friedman, M.: A comparison of alternative tests of significance for the problem of m rankings. Ann. Math. Stat. **11**(1), 86–92 (1940). https://doi.org/10.1214/aoms/1177731944. http://projecteuclid.org/euclid.aoms/1177731944
11. Garcia, C.: A metaheuristic algorithm for project selection and scheduling with due windows and limited inventory capacity. Kybernetes **43**(9), 1483–1499 (2014). https://doi.org/10.1108/K-11-2013-0245
12. Ghasemzadeh, F., Archer, N., Iyogun, P.: A zero-one model for project portfolio selection and scheduling. J. Oper. Res. Soc. **50**(7), 745–755 (1999)
13. Goldschmidt, O., Nehme, D., Yu, G.: Note: On the set-union knapsack problem. Nav. Res. Logist. **41**(6), 833–842 (1994). https://doi.org/10.1002/1520-6750(199410)41:6%3c833::AID-NAV3220410611%3e3.0.CO;2-Q
14. Gonçalves, J.F., Resende, M.G.: Biased random-key genetic algorithms for combinatorial optimization. J. Heuristics **17**(5), 487–525 (2011). https://doi.org/10.1007/s10732-010-9143-1
15. Gonçalves, J.F., Resende, M.G., Toso, R.F.: An experimental comparison of biased and unbiased random-key genetic algorithms. Pesqui. Oper. **34**(2), 143–164 (2014). https://doi.org/10.1590/0101-7438.2014.034.02.0143

16. Harrison, K.R., Elsayed, S., Garanovich, I., Weir, T., Galister, M., Boswell, S., Taylor, R., Sarker, R.: Portfolio optimization for defence applications. IEEE Access **8**(1), 60152–60178 (2020). https://doi.org/10.1109/ACCESS.2020.2983141. https://ieeexplore.ieee.org/document/9046777/
17. Harrison, K.R., Elsayed, S., Garanovich, I.L., Weir, T., Galister, M., Boswell, S., Taylor, R., Sarker, R.: A hybrid multi-population approach to the project portfolio selection and scheduling problem for future force design. IEEE Access **9**, 83410–83430 (2021). https://doi.org/10.1109/ACCESS.2021.3086070. https://ieeexplore.ieee.org/document/9446148/
18. Harrison, K.R., Elsayed, S., Sarker, R.A., Garanovich, I.L., Weir, T., Boswell, S.: Project Portfolio Selection with Defense Capability Options. In: GECCO 2021 - Companion Publication of the 2021 Genetic and Evolutionary Computation Conference, p. 1825–1826. ACM (2021). https://doi.org/10.1145/3449726.3463126
19. Harrison, K.R., Elsayed, S., Weir, T., Garanovich, I.L., Galister, M., Boswell, S., Taylor, R., Sarker, R.: Multi-period project selection and scheduling for defence capability-based planning. In: 2020 IEEE International Conference on Systems, Man, and Cybernetics (SMC), pp. 4044–4050. IEEE (2020). https://doi.org/10.1109/SMC42975.2020.9283334
20. Harrison, K.R., Elsayed, S., Weir, T., Garanovich, I.L., Taylor, R., Sarker, R.: An exploration of meta-heuristic approaches for the project portfolio selection and scheduling problem in a defence context. In: 2020 IEEE Symposium Series on Computational Intelligence (SSCI), pp. 1395–1402. IEEE (2020). https://doi.org/10.1109/SSCI47803.2020.9308608
21. Hifi, M., Michrafy, M., Sbihi, A.: Heuristic algorithms for the multiple-choice multidimensional knapsack problem. J. Oper. Res. Soc. **55**(12), 1323–1332 (2004). https://doi.org/10.1057/palgrave.jors.2601796
22. Holland, J.H.: Adaptation in natural and artificial systems: an introductory analysis with applications to biology, control, and artificial intelligence. MIT Press, Cambridge (1992)
23. Kellerer, H., Pferschy, U., Pisinger, D.: The multiple-choice knapsack problem. Knapsack Problems, pp. 317–347. Springer, Berlin (2004). https://doi.org/10.1007/978-3-540-24777-7_11
24. Kumar, M., Mittal, M.L., Soni, G., Joshi, D.: A tabu search algorithm for simultaneous selection and scheduling of projects. In: Yadav, N., Yadav, A., Bansal, J.C., Deep, K., Kim, J.H. (eds.) Harmony Search and Nature Inspired Optimization Algorithms. Advances in Intelligent Systems and Computing, vol. 741, pp. 1111–1121. Springer, Singapore (2019). https://doi.org/10.1007/978-981-13-0761-4_104
25. Liu, S.S., Wang, C.J.: Optimizing project selection and scheduling problems with time-dependent resource constraints. Autom. Constr. **20**(8), 1110–1119 (2011). https://doi.org/10.1016/j.autcon.2011.04.012. https://linkinghub.elsevier.com/retrieve/pii/S0926580511000665
26. Price, K.V.: An Introduction to Differential Evolution. New Ideas in Optimization, pp. 79–108. McGraw-Hill Ltd., New York (1999)
27. Puchinger, J., Raidl, G.R., Pferschy, U.: The multidimensional knapsack problem: structure and algorithms. INFORMS J. Comput. **22**(2), 250–265 (2010). https://doi.org/10.1287/ijoc.1090.0344
28. Shaffer, J.P.: Modified sequentially rejective multiple test procedures. J. Am. Stat. Assoc. **81**(395), 826 (1986). https://doi.org/10.2307/2289016
29. Shafi, K., Elsayed, S., Sarker, R., Ryan, M.: Scenario-based multi-period program optimization for capability-based planning using evolutionary algorithms. Appl. Soft Comput. **56**, 717–729 (2017). https://doi.org/10.1016/j.asoc.2016.07.009
30. Storn, R., Price, K.: Differential evolution-a simple and efficient heuristic for global optimization over continuous spaces. J. Global Optim. **11**, 341–359 (1997). https://doi.org/10.1023/A:1008202821328
31. Sun, H., Ma, T.: A packing-multiple-boxes model for R&D project selection and scheduling. Technovation **25**(11), 1355–1361 (2005). https://doi.org/10.1016/j.technovation.2004.07.010
32. Wei, Z., Hao, J.K.: Kernel based tabu search for the set-union knapsack problem. Expert. Syst. Appl. **165**(August 2020), 113802 (2021). https://doi.org/10.1016/j.eswa.2020.113802. https://linkinghub.elsevier.com/retrieve/pii/S0957417420306199

Analysis of New Approaches Used in Portfolio Optimization: A Systematic Literature Review

Danilo A. Milhomem and Maria José Pereira Dantas

Abstract This study aims to identify, by means of a systematic literature review, the main methods, tools, and techniques used for portfolio optimization, real-world constraints, and to analyze how the applications of this set were changing over time. The main difference of this research is the approach on several aspects that cover a variety of aspects associated with the portfolio optimization problem—the types of analyses, constraints, algorithms, and language/software used. The main results of the review show dynamic optimization as an emerging theme as well as the hybridization of algorithms and the search for constraints. This chapter presents the main realistic restrictions addressed in the works found, the algorithms of artificial bee colony and genetic programming, and the instances or markets where these algorithms were tested. A point of attention should be given to the input data of optimization models. Depending on the degree of the estimation error of these input parameters, the optimization results may be lower than the results of the 1/N trading strategy. Robust optimization, fuzzy logic, and prediction are examples of techniques used to reduce estimation errors. Regarding software/programming languages, Matlab appears as the most widely used and Python facilitates the task of programming and building models by people who are not specialists in programming. In addition, Python is also common due to its flexibility, object orientation, gratuity, and calculation speed.

1 Introduction

Portfolio optimization consists of determining a set of assets, and their respective portfolio participation weights, which satisfy the investor concerning the combination of risk-return binomial. Given this problem, Markowitz [55] proposed the mean-variance (MV) model. In his model, the expected return is given by the average of

Note: This chapter is an extended version of a previously published article: [61] (licensed under the Creative Commons Attribution License).

D. A. Milhomem (✉) · M. J. Pereira Dantas
Pontifícia Universidade Católica de Goiás (Catholic University of Goiás), Goiânia, GO, Brazil

K. R. Harrison et al., *Evolutionary and Memetic Computing for Project Portfolio Selection and Scheduling*, Adaptation, Learning, and Optimization 26,
https://doi.org/10.1007/978-3-030-88315-7_6

the historical data of the stock's return and the risk is calculated as the variance of these returns. The main idea of the MV model is to deal with the returns of individual assets as random variables and to adopt the value of expected return and variance in order to quantify the return and investment risk, respectively [94].

If the customer wants to minimize the risk for a given value of fixed return, then the mathematical model that will select the stocks is given by:

$$\min Var\,[\xi_1 x_1 + \xi_2 x_2 + \cdots + \xi_n x_n] \tag{1a}$$

subject to

$$E[\xi_1 x_1 + \xi_2 x_2 + \cdots + \xi_n x_n] \geq \alpha \tag{1b}$$

$$x_1 + x_2 + \cdots + x_n = 1 \tag{1c}$$

$$x_i \geq 0, \ \ i = 1, 2, \ldots, n,$$

where E represents the operation of expected value, Var is the variance operator, x_i is the proportion of the total amount available for investment applied in the stock i, and ξ_i represents the random return for the ith stock [94]. Note that, for a given level of risk, the maximization of returns is also an equivalent model. The objective function of minimizing the risk is given by Eq. (1a). In Eq. (1b), α represents the minimum return value that the investor can accept. The budget constraint is given by Eq. (1c) and indicates that all available money must be invested. Finally, the non-negativity constraint means that selling short is not accepted.

Merton [60] approaches the portfolio optimization problem from the perspective of how to systematically build and analyze optimal dynamic models of continuous-time under uncertainty. According to Merton, the main advantage of the continuous-time model compared to models that consider discrete time is the fact that the continuous-time model contemplates only two types of stochastic processes: functions of Brownian movements and Poisson processes. Thus, the number of parameters in the problem is reduced and allows taking full advantage of the huge amount of written literature about these processes.

In a study by Uryasev [85], conditional-value-at-risk (CVaR) is proposed as a risk measure of the portfolio and the optimization occurs relative to this metric. The CVaR is derived from the weighted average of "extreme" losses in the tail of the distribution of possible returns in addition to the cutoff point of the value-at-risk (VaR). One of the advantages of CVaR is that it can be minimized efficiently using linear programming (LP) techniques.

Another approach to the portfolio problem is robust optimization, which, according to Fabozzi et al. [30], is the act of incorporating into the model, through mathematical techniques, the uncertainty generated by the errors of estimation of the parameters used in optimization. According to Pflug and Wozabal [70] and Fabozzi et al. [30], robust optimization offers an advantage due to the fact that modifications in the mathematical model do not change the characteristics of the problem, which remains a problem of quadratic programming.

DeMiguel et al. [25], Pflug et al. [71], and Behr et al. [14] report that the uniform investment strategy (naive diversification or 1/N) is more rational the greater the degree of uncertainty about the risk/return distribution is. This fact is supported by several experimental studies that demonstrate that this kind of strategy tends to be better in environments of extreme uncertainty and is usually a good strategy for risk-averse investors. DeMiguel et al. [25] show in what situations naive diversification is preferred to optimization. The results by Behr et al. [14] corroborate the results of DeMiguel et al. [25], showing that the optimization strategies are not unanimously better than the 1/N strategy from the perspective of the estimation error outside the sample.

Sparse and stable optimization is an important advancement in mean-variance portfolio selection. In general, DeMiguel et al. [24], Brodie et al. [18], Fan et al. [31], Behr et al. [14], and Dai and Wen [23] defend the idea of taking into account weights that can assume either positive or negative values (i.e., short and long positions). The idea is to stabilize the weights of the assets to reduce the estimation error using the mean-variance model with the addition of a restriction that the norm of the portfolio weight vector must be less than a certain limit. The sum of the weights can be less than or equal to 1 (1-norm). In addition, the model consists of admitting negative values (short) but normalization brings them to the positive side. Furthermore, the threshold of the sum of the weights can be greater than or equal to 1. This simple change brought better results in terms of Sharpe ratio than strategies present in the literature.

Some authors have conducted reviews on portfolio optimization methods. For example: Hu et al. [40] studied the application of evolutionary computation in the discovery of rules in algorithmic trading for shares; Ertenlice and Kalayci [29] conducted a review on swarm intelligence research for portfolio optimization, discussing algorithms and applications; Zhang et al. [94] reviewed the portfolio optimization problem with the Markowitz mean-variance structure; and Mansini et al. [54] examined over 20 years of portfolio optimization based on operational research.

The contribution of this chapter lies in the fact that no article has been found in the literature that broadly and jointly encompasses all of the following aspects: both heuristic and exact methods, realistic constraints, the evaluated markets, the types of stock analysis, and the software/programming languages employed by the considered studies.

Given the importance of portfolio optimization and the scope of related matters, this study has as purpose to identify by means of a systematic literature review (SLR) the main methods, tools, and techniques of portfolio optimization, their real-world constraints, and to analyze how the applications of this set were changing over the years.

The remainder of this chapter is organized as follows: Sect. 2 presents the method through which state of the art was sought in the literature on portfolio optimization. In Sect. 3, the main results of the review are presented. In Sect. 4, the results are discussed and the research questions are answered. Finally, the conclusions are shown in Sect. 5 with Sect. 6 discussing the research gaps that were identified through this study.

2 Research Method

This study is categorized as a configurative review as it explored the theme defined with qualitative data and data gathered from more heterogeneous primary studies. The SLR steps performed were based on Dresch et al. [26] and complemented according to the studies of Bastos et al. [13], Furlan and Laurindo [33], and Yu et al. [90]. The stages of the chapter's development are documented in the protocol for systematic literature review, shown in Table 1.

2.1 Research Questions

This section describes the approaches to each review question given in Table 1. Regarding Question 1, the objective is to find the main optimization models developed in papers with high quality standards and produced in the literature between 2014 and 2018. In Question 2, the aim is to find what the authors have written about the constraints and what impact they have on the model in terms of difficulty and complexity. In Question 3, the purpose is to find what type of approach on technical and fundamental analysis the authors have used in their papers and the percentage of participation for each type of analysis. Question 4 has the objective of discovering which software or programming languages are most commonly used to solve these problems.

The research questions in Table 1 guided the authors on the aspects that should be observed in reading the compiled articles. Discussion and answers to these research questions are provided in Sect. 4.

2.2 Search Sources

The research process performed is a specific set of papers published between 2014 and 2018, both inclusive. The database chosen was Web of Science because it is comprehensive and strict in the evaluation process. The journals were selected through the criteria established in the SLR protocol given in Table 1. The process of identifying the impact factor in each journal was manual with the aid of the Web of Science website.

Five searches were performed with the respective search terms shown in Table 1. The choice of these terms was made to enable the search to bring the greatest number of papers within the theme addressed in the research.

Table 1 Protocol for systematic literature review

Central theme: The central theme is portfolio optimization. This theme will be approached in an open manner with the objective of finding the greatest number of academic and scientific publications	
Conceptual framework: Portfolio optimization is a very important topic in the stock market area. Since the creation of the mean-variance model by Markowitz [55], several studies have been developed in order to improve the model or create more robust analysis methods through the insertion of realistic constraints in the model. This review has the purpose of identifying the studies that include the theme, the tools, and methods for optimizing portfolios	
Background	Diversification and optimization of an investment portfolio
Scope	The search period will be between 2014 and 2018, with the intention of finding the most relevant publications published in these five years
Theoretical currents	Portfolio optimization
Languages	Both the search terms and the papers will be in English
Research questions	**Question 1**—Which key methods, tools, or techniques are used in portfolio optimization?
	Question 2—Which realistic constraints are used?
	Question 3—What type of analysis is done regarding the stock: fundamental, technical, or mixed (fundamental and technical)?
	Question 4—Which software/programming languages are used?
Review strategy	(✓) Configurative ( ) Aggregative
Inclusion criteria	The papers should contain the latest discoveries about portfolio optimization models as well as realistic constraints, analysis types, and algorithms
Exclusion criteria	Duplicate papers
Search terms	**1st search**: "optimization" AND "portfolio" AND "algorithms" AND "return" AND "risk";
	2nd search: "optimization" AND "portfolio" AND "algorithms" AND "return" AND "risk" AND "stock market";
	3rd search: "portfolio optimization" AND "stock market" AND "Markowitz";
	4th search: "stock market" AND "portfolio selection" AND "portfolio optimization";
	5th search: "portfolio optimization" AND "stock market" AND "financial"
Search sources	Web of science database

2.3 *Inclusion Criteria and Exclusion Criteria*

The papers selected were those published in journals in the period between 2014 and 2018. Other than this criterion, Fig. 1 has the inclusion or exclusion filters for the papers. The papers were selected according to their approach regarding the portfolio optimization theme and their potential relevance to addressing the research questions in Table 1.

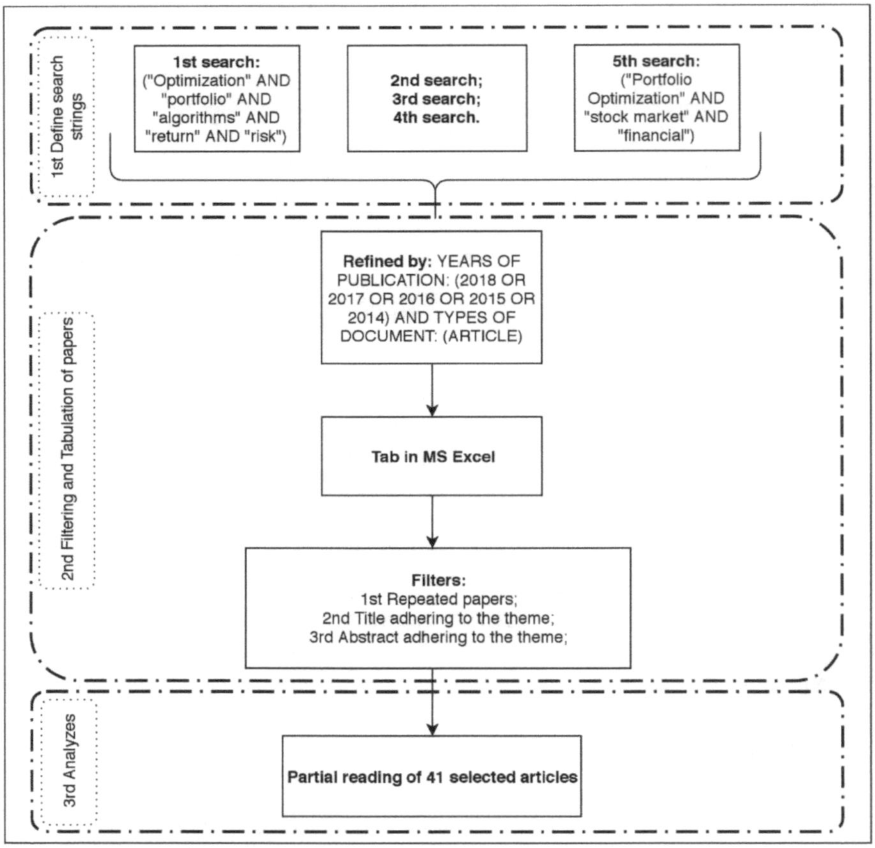

Fig. 1 Inclusion criteria and exclusion criteria

2.4 Data Extraction

The data extracted from the selected papers were:

- The source (journal) and complete reference;
- Research areas;
- The authors, their respective institutions, and the country where the published journal is located;
- The study abstract including the main research questions and the answers;
- Research questions/objective;
- The methodology and mathematical model;
- Database (instances or platforms with stock prices);
- Software/programming language;
- Types of constraints implemented in the model.

A partial reading of the papers was performed to extract the information cited above and the data tabulation was made with the aid of spreadsheets in Excel.

2.5 Data Analysis

The data were tabulated to show:

- The number of paper published per year and the country of origin;
- If the papers present real-world constraints, also known as realistic constraints (addressed in Question 2);
- The number of studies in each main category: Exact methods, heuristic, software and programming language (regarding Question 1, Question 4);
- The paper's approach regarding the stocks evaluation and their respective companies. In other words, the type of analysis made using the data (addressing Question 3).

2.6 Deviations in the Protocol

Some research steps were changed regarding what was proposed:

- The use of the impact factor in journal evaluations instead of the Qualis classification;
- Inclusion of analysis on the data used to test the algorithms or models (whether they were real or of instances);
- Inclusion of VosViewer software for macro analysis of the articles found and their references to classical authors;

Table 2 Qualifications of journals

Journal	Impact factor	Papers
IEEE Transactions on Fuzzy Systems	12.029	2
Swarm and Evolutionary Computation	7.177	5
Energy	7.147	1
Energy Economics	7.042	1
Expert Systems with Applications	6.954	12
Information Sciences	6.795	1
Applied Soft Computing (print)	6.725	2
Neurocomputing	5.719	1
Neural Computing and Applications	5.606	2
Ecological Economics	5.389	1
European Journal of Operational Research	5.334	5
IEEE Transactions on Signal Processing	4.931	1
Annals of Operations Research	4.854	3
Management Science	4.883	2
International Journal of Fuzzy Systems	4.673	1
Science China—Information Sciences	4.380	1
Financial Innovation	3.985	1
International Journal of Approximate Reasoning	3.816	1
Soft Computing	3.643	2
Computational Intelligence and Neuroscience	3.633	1
Electric Power Systems Research	3.414	1
International Journal of Environmental Research and Public Health	3.390	1
IEEE Access	3.367	3
Physica A—Statistical Mechanics and its Applications	3.263	6
Economic Modelling	3.127	4
Journal of the Operational Research Society	2.860	1
Operations Research	2.410	1
Central European Journal of Operations Research	2.345	1
Journal of Optimization Theory and Applications	2.249	1
Journal of Statistical Mechanics—Theory and Experiment	2.232	1
International Journal of Information Technology & Decision Making	2.220	1
OR Spectrum	1.652	1
Quantitative Finance (print)	1.491	2
Mathematical Problems in Engineering (print)	1.305	1
PLOS One*	N/D	1
Intelligent Systems in Accounting, Finance, and Management	N/D	1

* PLOS One does not consider the Impact Factor

- To update the systematic literature review, a section was created to place the results of articles found between 2019 and 2021. The structure of this topic has a different configuration as it is a deviation, although necessary. The periodicals of the new search were classified in Table 2.

3 Results

The following is a macro view of the results of the search for papers on the portfolio optimization theme. By combining search terms into a single set of search strings ("optimization" AND "portfolio" OR "portfolio optimization" OR "portfolio selection" AND "algorithms" AND "return" AND "risk" AND "stock market" AND "Markowitz" AND "financial") and refining for works written in the English language, 2137 papers on portfolio optimization were found. With these papers, some analyses were performed with the aid of VosViewer software. A survey of the keywords used in the titles of the articles was performed, the mapping of which can be seen in Fig. 2.

Note in Fig. 2, the presence of terms such as "transaction costs", "risk management", "dynamic programming", and "robust optimization". Besides, the following terms related to portfolio optimization have also been identified: "genetic algorithms", "multi-objective optimization", and "stochastic optimization". These terms give clues about the lines of research explored by the articles found.

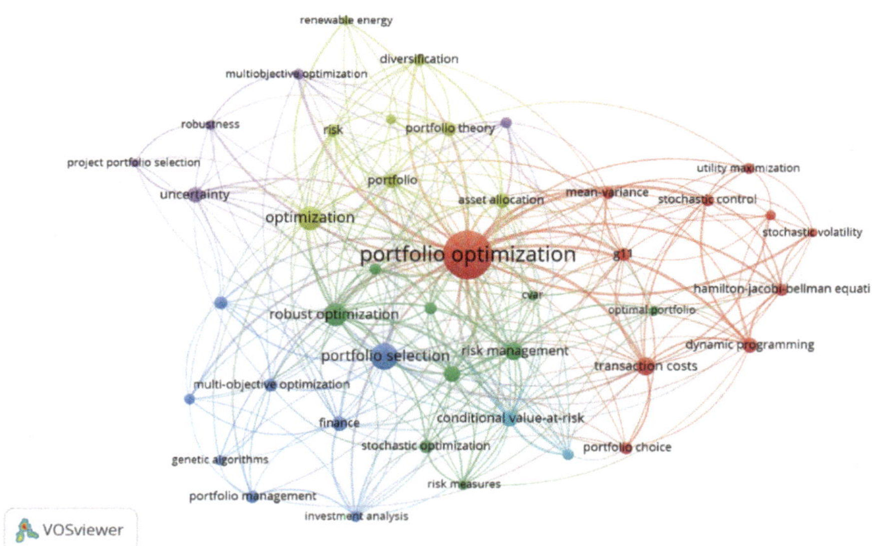

Fig. 2 Map of keywords contained in the title of the publications

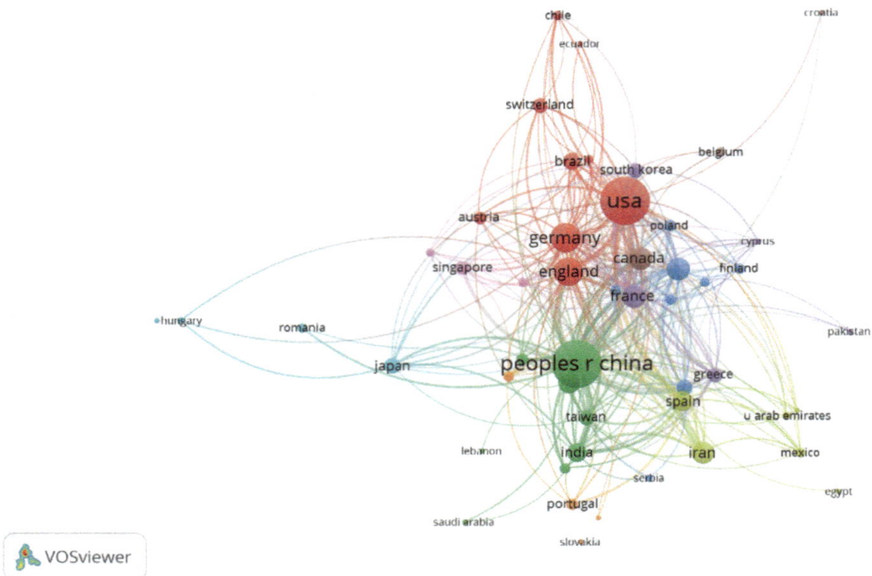

Fig. 3 Country mapping in terms of publication

In Fig. 3, the main countries where the papers were published are shown. China and the USA appear as the two countries that publish the most on the topic under study, followed by Germany, France, Canada, and England. Brazil also appears on the map with a considerable number of publications on the subject.

In Fig. 4, the mapping of the most cited authors in the studies researched is shown. Each author is identified on the map along with the year of publication of the aforementioned work and the name of the journal.

Markowitz [55] appears as the most cited author, evidencing that the mean-variance model is still widely discussed in the financial market and the universe of portfolio optimization. In addition to Markowitz, authors such as Zadeh [91], DeMiguel et al. [25], Uryasev [85], and Merton [60] wrote important articles on the topic of portfolio optimization and all of these authors continue to be widely cited.

3.1 Journal Impact Factor

Table 2 shows the journals where the analyzed papers were published. This table presents two important pieces of information: the impact factor (an internationally recognized index used to measure the journal's relevance and scope in publications)

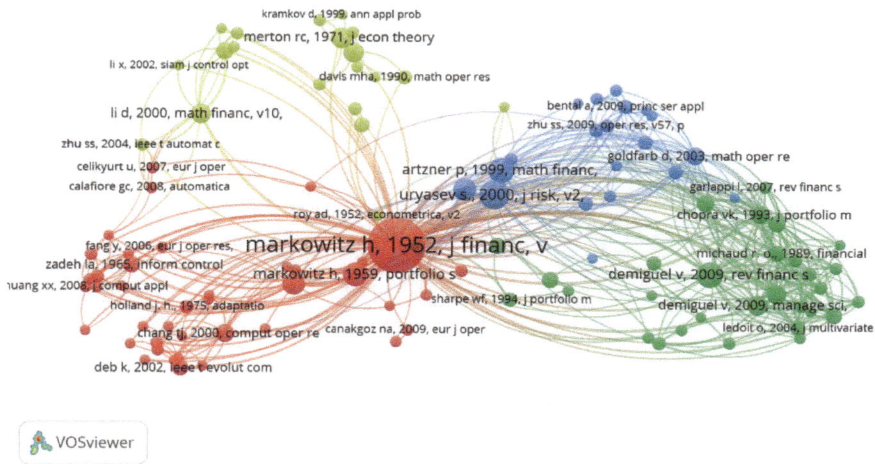

Fig. 4 Mapping of the most cited authors

and the number of papers per journal. The data in Table 2 shows that the selected papers come from excellent journals.

3.2 *Classification of Methods*

As a result of the search process, a total of 73 unique studies (without duplicity) were identified after the application of the searches and the inclusion and exclusion criteria/filters. In Tables 3 and 4, only the papers on which the partial reading was carried out, with the exception of the two bibliographic review papers ([29] and [54]), are shown. That is, only 71 studies are presented. Among the 71 papers, 39 studies published between 2014 and 2018 were used to classify according to the method, algorithms, software, and constraints/restrictions. Furthermore, 32 papers published between 2019 and 2021 were used to show the latest in portfolio optimization research. Table 3 presents the list of papers that used exact methods to solve the portfolio optimization problem, evidencing the strong participation of these methods for the selection of the best assets (stocks).

Table 3 Papers that used exact methods

References	Technique: Exact
[4]	LVaR
[9]	DR framework
[2]	MV
[65]	MINLP
[6]	MV
[12]	PBR
[15]	MedMAD
[16]	Markov chains
[19]	MILP
[41]	MinG
[48]	SSDMN
[56]	IMPRO
[75]	MST
[80]	Mean-variance-asymmetry-kurtosis
[82]	DCCA
[83]	MAD
[96]	q-dependent
[97]	Temporal network structure

MAD: mean-absolute deviation model; **PBR**: performance-based regularization-combinatorial optimization; **MV**: mean-variance; **MINLP**: mixed-integer nonlinear programming; **MedMAD**: median mean absolute deviation (MAD); **MILP**: mixed-integer linear programming; **IMPRO**: insured multi-period portfolio robust optimization—linear model; **MinG**: minimal Gini model; **LVar**: liquidity adjusted value at risk; **DR*Framework***: extension of the Markowitz model where the proxy variance was substituted for the downside risk as asymmetric lower partial moments (ALPM); **SSDMN**: stochastic domain for mixed normal distributions; **DCCA**: method for cross–correlation analysis; **q-dependent**: q-dependent cross-correlation coefficient; **MST**: dynamic portfolio strategy based on network structures (minimum spanning tree)

Most of the studies presented in Table 4, which use heuristics, addressed multi-period and multi-objective problems. Fuzzy logic, genetic algorithms (GAs), and swarm intelligence algorithms (emphasis on particle swarm optimization—PSO) are the methods that appear most often in the studies.

According to Table 4, among the articles that use heuristics 23.8% use hybrid algorithms, 38.09% consider a single-period, 47.61% address multi-period problems, and 71.4% consider more than one objective function (i.e., multi-objective problems). Figure 5 shows the distribution of papers according to the year of publication.

Table 4 Papers that used heuristic methods

References	Technique: Exact/Heuristic	Heuristic	Hybrid heuristic	Single-period	Multi-period	Single-objective	Multi-objective
[10]	MOPSO	X		X			X
[17]	GP	X			X		X
[21]	MOPSO	X			X		X
[35]	GA and TS	X			X	X	
[46]	M-CABC	X		X			X
[49]	Immunological algorithm	X		X			X
[51]	GA with Fuzzy		X		X		X
[53]	NSGA-II and SPEA 2	X			X		X
[58]	NSGA-II; MOEA/D and GWASF-GA	X			X		X
[62]	SR-MOPSO	X		X			X
[66]	MODE/Fuzzy and MOES/Fuzzy		X		X		X
[69]	Algorithm of differential evolution	X		X		X	
[73]	(NMOEA/D)	X			X		X
[76]	GA	X		N/D	N/D	X	
[72]	IWO and PSO	X		X		X	
[77]	FTS	X					
[78]	FLR	X					
[79]	MDRS		X	X			X
[81]	GA	X			X		X
[86]	GA with Fuzzy		X	X			X
[93]	GA with Fuzzy		X		X		X

GA: genetic algorithm; **Fuzzy**: fuzzy programming; **MPSO**: multi-purpose particle swarm optimization algorithms; **TS**: tabu search; **NSGA-II**: non-dominated sorting genetic algorithm II; **SPEA 2**: strength Pareto evolutionary algorithm 2; **GP**: genetic programming; **NMOEA/D**: normalized multi-objective evolutionary algorithm based on decomposition; **MODE/Fuzzy**: multi-objective differential evolution/fuzzy; **MOES/Fuzzy**: multi-objective evolution strategy/fuzzy; **GWASF-GA**: global weighting achievement scalarizing function genetic algorithm; **M-CABC**: artificial bee colony algorithm based on multi-objective covariance; **SR-MOPSO**: self-regulating multi-objective particle swarm optimization; **IWO**: invasive weed optimization; **PSO**: particle swarm optimization; **FTS**: methods of weighted diffuse time series: **FLR**: fuzzy logic relation; **TE**: the difference between the tracking portfolio return and the benchmark return; **P/E**: P represents the market value of each asset and E represents the profit for each asset; **MDRS**: mean downside risk-skewness

4 Discussion

4.1 Which Key Methods, Tools, or Optimization Techniques Are Used in the Portfolio Optimization Problem?

In 18 studies, the authors use exact techniques to solve the portfolio optimization problem starting from the MV model or by adapting it (i.e., inserting new parameters instead of the mean and variance) to what some authors call the extension of

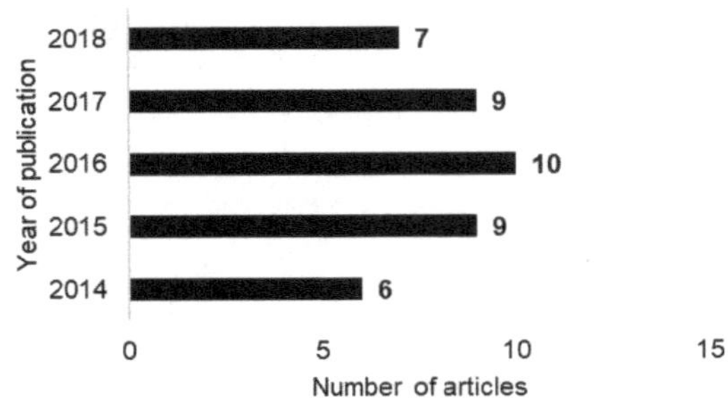

Fig. 5 Papers distribution per year of publication

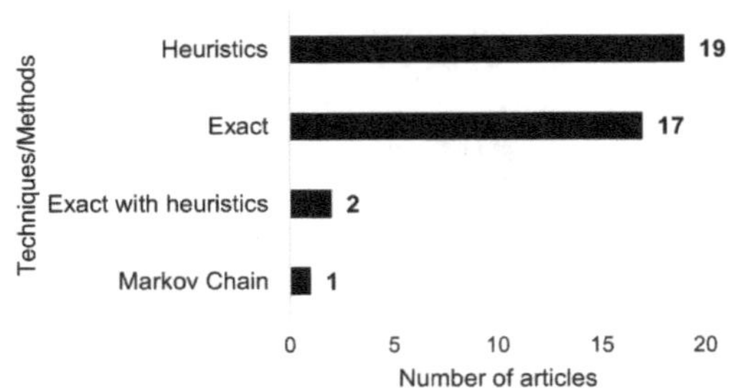

Fig. 6 Papers distribution per techniques/methods for portfolio optimization

Markowitz model [9, 10, 12, 16, 19, 40, 46, 48, 51, 56, 65, 72, 77, 80, 83, 93, 96, 97].

Despite the evolution of new techniques to solve portfolio optimization problems (i.e., stochastic/heuristic techniques), many investors and researchers prefer exact techniques due to their characteristic ability to find the global optima. Regardless of the degree of difficulty of modeling or computational spending that will be required, exact techniques always return the optimal solution so they are so attractive, especially for conservative investors.

The number of published studies that used heuristics to solve the portfolio optimization problem demonstrates the strong participation of these methods [10, 17, 21, 29, 35, 46, 49, 51, 53, 58, 62, 66, 69, 72, 73, 76, 79, 81, 86]. Only one of the selected studies shows the use of Markov chains [16]. In Fig. 6, the distribution of the resolution techniques for the portfolio optimization problem is shown. It is possible to notice the strong participation of heuristics solutions, although the exact techniques have an extensive presence.

One of the justifications for the use of metaheuristics is the fact that they generally find a solution close to optimal with a lower computational expense than exact methods. Furthermore, depending on the degree of complexity of the model, these techniques converge to the optimal solution faster than the exact ones. Another advantage of heuristics is the ease of working with multi-objective problems and with several restrictions.

According to Table 4, about 71% of articles that use heuristics consider multi-objective portfolio optimization. Multi-objective problems are solved either as a single-objective problem in which the objective function is a weighted sum of all the objective functions, optimizing one objective (most important one) only by converting other objectives as constraints, optimizing a different objective function such as deviation minimization from the possible goal of each objective (goal programming approach), or optimizing all the objectives simultaneously (which is the case in evolutionary and population based approaches such as NSGA-II). Example objectives include maximizing the return and minimizing various measures such as the risk, transaction costs, degree of model uncertainty, value-at-risk, or conditional value-at-risk. Hybrid algorithms composed of fuzzy programming, GA, MOES, and NSGA-II have been widely used to solve problems with more than one objective function.

The main goal of hybridization is to unite the advantages of two or more different algorithms and to build a more robust model. For example, Zhang and Liu [93] propose a hybrid model using fuzzy programming and a GA . With fuzzy programming, it is possible to solve multi-objective problems by transforming it to a single-objective problem. In addition, expert knowledge can be inserted in the portfolio selection model (optimization) using fuzzy [51, 77, 93].

Attention should be drawn to two particular techniques in the development of new studies: artificial bee colony (ABC) and genetic programming (GP). The ABC algorithm has proven efficient and stands out by allowing the resolution of optimization problems with realistic constraints. Other important features are the exploration cycles (exploration of global optima) and continuous exploitation (exploitation of local optima) that enable a wide exploitation of the optimal values. The ABC algorithm, when compared to other population and stochastic algorithms of the same class, presents similar or better performance. The other advantage of ABC is that it uses fewer control parameters than other algorithms [29, 46].

A robust GP approach was presented to automate the purchase and selling decisions in order to maximize the Sterling ratio (i.e., the total return divided by the maximum drawdown). The method is the RSFGP—a method of genetic programming by random sampling. It was the only study found that presents a robust approach to daily operations on the stock exchange. The RSFGP works as follows: in the initial evaluation period, the portfolio starts with a certain amount W_0 of cash. In each day, the GP rule is evaluated and the stocks classified as "True" are purchased (or retained if they are already in the portfolio) and those classified as "False" are sold if they belong to the portfolio. The total value of the W_t portfolio is calculated daily by the stocks evaluation at the closing price of each day plus the cash portfolio value. The great advantage of this model is the fact that it allows the management of a stock portfolio in an automatic way, without requiring the help of financial market specialists and without having to make constant adjustments in the model [17].

The choice of method to solve the optimization problem depends primarily on what the researcher wants to obtain concerning the expected results:

- The complexity of the model—difficult models are commonly solved by heuristics;
- The time spent to get a solution. More powerful methods, like the exact ones, spend more time and deliver optimal solutions while heuristic methods spend less time but deliver a n approximate solution that can be considered feasible.

Furthermore to these approaches, the investor can choose the backtesting method where he can verify the efficiency of the entire strategy in the most varied combinations between mathematical models, exact and heuristic methods, restrictions, and 1/N strategy [25]. Generally, the decision-making between using 1/N in the opposite of optimization is based on metrics used to assess the degree of uncertainty of the model. If this measure is high, the 1/N strategy is preferred.

DeMiguel et al. [25] report that the portfolio optimization strategy has advantages over benchmark 1/N in three situations: (1) if there is an abundance of historical data available; (2) if the Sharpe ex-ante value (expected) of the efficient medium variance portfolio is substantially higher than the Sharpe value of the 1/N portfolio; and (3) the number of assets that will make up the portfolio is small. The first two conditions are intuitive and the reason for the latter condition is that a smaller number of assets implies fewer parameters to be estimated and, therefore, less space for estimation error. This makes 1/N diversification less effective than diversification by optimization.

4.2 Which Realistic Constraints Are Used?

Constraints make the model more complex to be solved. However, they also make the portfolio optimization process much easier. Table 5 shows the constraints, the authors who cited them, and the method used by the authors to solve the problems. Constraints on non-negativity and that the total value available must be fully invested have not been considered as they are considered trivial in portfolio optimization models. There is an extension to comment on the databases that the authors used to test their algorithms in this section. In some studies, the OR-Library instance was used, which allows free access to the data. The data sources are well diversified and among the papers, only one used data from Brazil through the Thomson-Reuters Datastream platform [53].

With several studies that succeeded the model proposed by Markowitz [55], some new constraints were proposed such as the cardinality constraints and minimum purchase limits constraints. These constraints provide the model with a non-convex search space, which makes the problem difficult, in terms of computational cost and response time, to be solved using exact approaches [17, 29, 35, 46, 58, 66, 72, 73, 94]. A description of the meaning of each constraint shown in Table 5 is given as follows:

- Transaction costs: in financial markets, there are transaction costs arising from the process of purchasing and selling bonds and investment portfolio reviews;
- Transaction lots—rounds: the minimum transaction unit required to invest in an asset;
- Cardinality constraint: restricts the number of bonds that will compose the investment portfolio;

Table 5 Realistic constraints

References	Heuristic method	Realistic constraints	Database
[10]	MOPSO	Threshold constraints; Cardinality constraints;	S&P500
[17]	GP	N/D	Stock Market Espanha (IBEX35 index)
[21]	MOPSO, NSGA-II and SPEA 2	Long-only constraint	Stock Market China
[35]	GA and TS	Cardinality constraints	OR-Library
[46]	M-CABC	Cardinality constraints; threshold constraints; long only constraint;	OR-Library
[49]	Immunological algorithm	Cardinality constraints; threshold constraints; transaction cost;	Indices: Hang Seng 33, DAX 200, FTSE 100, S & P 100 and Nikkei 225
[51]	GA with Fuzzy	Cardinality constraints; transaction cost; transaction lots;	Stock market China
[53]	NSGA-II and SPEA 2	Cardinality constraints; transaction cost; transaction lots;	Thompson-Reuters Datastream (Now Eikon)
[58]	NSGA-II, MOEA/D, and GWASF-GA	Cardinality constraints; quantity constraint; self-financing constraint; transaction cost.	Fama and French Data Library
[62]	SR-MOPSO	Cardinality constraints; threshold constraints;	OR-Library
[66]	MODE/fuzzy and MOES/fuzzy	Long-only constraint	BSE Index (Bombay Stock Exchange, India)
[69]	Algorithm of differential evolution	Non-constrained problem with continuous variables	Dow Jones Index
[72]	IWO and PSO	Sector capitalization; transaction lots; cardinality constraints; threshold constraints;	Tehran Stock Exchange Market in 2013
[73]	NMOEA/D	N/D	Stock Market China
[76]	GA	Tracking error constraints.	US bond index
[79]	MDRS	Cardinality constraints	Stock Market Espanha (IBEX35 index)
[81]	GA	Cardinality constraint; quantity constraint; long only constraint; transaction costs.	S&P500
[86]	GA with Fuzzy	Cardinality constraints	Stock Market Espanha (IBEX35 index)
[93]	GA with Fuzzy	Self-financing constraint; the long only constraint	Stock Market Shanghai

GA: genetic algorithm; **Fuzzy**: fuzzy programming; **MPSO**: multi-purpose particle swarm optimization algorithms; **TS**: tabu search; **NSGA-II**: non-dominated sorting genetic algorithm II; **SPEA 2**: strength Pareto evolutionary algorithm 2; **GP**: genetic programming; **NMOEA/D**: normalized multi-objective evolutionary algorithm based on decomposition; **MODE/Fuzzy**: multi-objective differential evolution/fuzzy; **MOES/Fuzzy**: multi-objective evolution strategy/fuzzy; **GWASF-GA**: global weighting achievement scalarizing function genetic algorithm; **M-CABC**: artificial bee colony algorithm based on multi-objective covariance; **SR-MOPSO**: self-regulating multi-objective particle swarm optimization; **IWO**: invasive weed optimization; **PSO**: particle swarm optimization; **FTS**: methods of weighted diffuse time series: **FLR**: fuzzy logic relation; **TE**: the difference between the tracking portfolio return and the benchmark return; **P/E**: P represents the market value of each asset and E represents the profit for each asset; **MDRS**: mean downside risk-skewness

- Investment threshold constraints: these constraints define upper limits of the ratio-amount of each asset held in the portfolio;
- Decision dependency constraints: decision dependency requirements that are common in financial transactions;
- Quantity constraint: the quantity constraint is used to limit the position size by placing maximum and minimum values. The minimum limit is set to 5% of the portfolio value at the time of the transaction and the maximum value is 20%. The lower limit avoids virtually insignificant positions for the portfolio performance and the upper limit averts too much exposure or weight for any stock;
- Long-only constraint: means that the realization of selling short operations is not permitted, i.e., the weight invested in any stock is always positive;
- Self-financing constraint: no money is withdrawn or added during the operation.

4.3 *What Type of Analysis Is Done Regarding the Stock: Fundamental, Technical, or Mixed (Fundamental and Technical)?*

There are few studies using fundamental analysis [2, 6, 51, 76]. One of the rationales for this lack of studies is the complexity in analyzing the companies' financial performance data; investors who use fundamental analysis are more conservative and invest in long term. The technical analysis is generally simpler, since it analyzes few variables, and has an advantage in terms of the analysis speed in daily operations, for example.

Most papers use technical analysis [4, 12, 15, 16, 19, 46, 48, 49, 53, 56, 62, 65, 73, 77, 80, 82, 83, 93] because this is easy to access in instances and/or from stockbrokers. In addition, investors who use this data want to get rapid gains in daily, weekly, or monthly operations, identified through simpler analyses and fewer variables, the opportunities offered by the market. Figure 7 presents the distribution of the types of analysis in the selected papers.

Silva et al. [81] propose a mixed approach of the two types of analysis and demonstrate through simulations that the stock optimization based on financial indices,

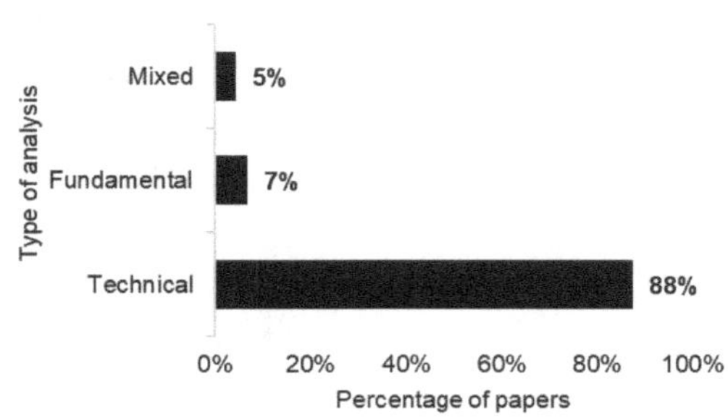

Fig. 7 Papers distribution per type of analysis

arising from fundamental analysis, can be used to choose the best companies in operational terms, thus obtaining returns above market average with low variation in their returns. The financial indices that enable the company's fundamental analysis in terms of profitability, liquidity, debt, and growth are provided here [81]:

- Debt index (DR): used to measure the company's indebtedness level;
- Net return (ROE): measures financial performance in the case of generating profits using company assets;
- Profit margin index (PM): assesses the company profitability by calculating the percentage of retained earnings after paying the operational, administrative, and financial costs in addition to taxes;
- Price-profit index (PER): a valuation ratio of the current price of a company's share compared to its earnings per share, used to choose companies that are less valued in the market;
- Revenue growth percentage (GR): an economic indicator that shows the business evolution;
- Positive trend of net profit (NI): shows the increase in profits and aims to show the companies with the highest growth;
- Common shares in circulation (CSO): represents the fundamental owner of the corporation by shareholders. When the company issues shares, this number is added to the previous value in the balance sheet and represents an increase in the total number of shares. Hence, the value of the company is distributed among a larger number of shares.

The following are the parameters used for a technical negotiation, as given by Silva et al. [81]:

- The global value of each stock: a daily assessment of the quarterly financial indexes and the daily PER, which is updated using the closing price to classify the algorithm actions to make a business decision;
- Stop loss and protection of capital: predetermined policies that reduce the portfolio exposure leaving a losing position, not allowing one or more investments to continue to lose money, and protect the remaining capital available to continue the investment activity;
- Take profit: percentage of pre-established profit. When it is reached, the stock is sold;
- Position size: the percentage of the current value of the portfolio to invest in each new stock. This defines the level of concentration or diversification in the portfolio.

4.4 Which Software/Programming Languages Are Used?

Among the explored studies, 26 did not mention the software used and only provided the pseudocode or the mathematical model. Here, the software refers to a method for solving a given mathematical model and may be either an established software with

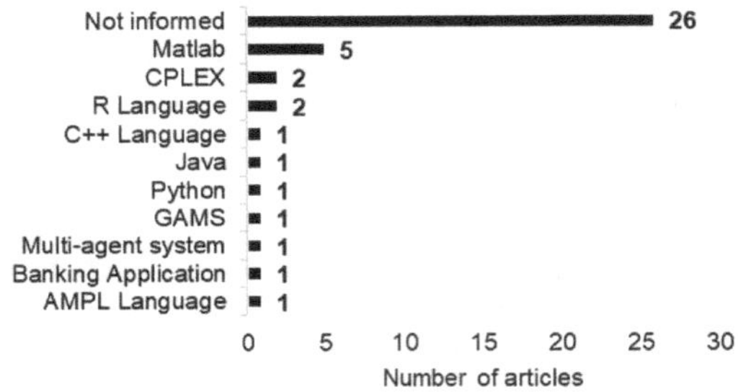

Fig. 8 Papers distribution per software/programming language

a user interface (which may require input preparation) or a programming language that the authors used to implement the proposed model themselves.

The distribution of software or programming language usage is given by Fig. 8. There are tools with all kinds of features such as free tools, paid tools, tools that are easy to learn and that have a variety of libraries/packages available that can be used by researchers who want to solve a problem without necessarily learning a programming language.

The R programming language offers some advantages such as the best learning curve, a wide variety of packages ready for optimization or data science, and active forums for discussion and learning. Despite requiring certain programming knowledge, R appears as a nice option for portfolio optimization scholars as it has an easier learning curve than other languages.

According to Karakalidis and Sifaleras [44], AMPL is an easy-to-learn language where the researcher implements the mathematical model using a natural language, with notations very close to those used in the writing of the model, and allows the user to execute optimization software. Among these solvers, we highlight CPLEX®, GUROBI Optimizer®, and KNITRO® (used for nonlinear problems). The choice of solver depends on the type of problem to be solved (e.g., linear programming, quadratic programming, mixed-integer quadratic, or nonlinear programming). These solvers often provide a free version for academic research. However, these academic versions may be limited and depending on the size/complexity of the problem, it may be necessary to acquire more complete, paid versions of the software.

The Python programming language has interesting features that enable the construction of a flexible program or algorithm. In addition, object-oriented programming, materials available on the Internet, and calculation speed are other advantages of this programming language [17, 37, 57, 63]. In addition to the libraries, there is a module called Scikit-Learn that integrates a wide range of state-of-the-art machine learning algorithms for supervised and unsupervised mid-scale issues. This package focuses on making machine learning accessible to non-specialists by using a high-level, general-purpose language [68].

Despite some differences, the category of languages composed of Python, R, C++, and Java all require certain knowledge about programming, regardless of having libraries/packages that facilitate ease of use and their free availability. Additionally, due to their flexibility, these languages are better for researchers or investors who want to work with heuristics to solve optimization problems.

4.5 Recent Researches

In the previous sections, the research questions were answered according to the first stage of the review (i.e., papers published between 2014 and 2018). In this subsection, recent work from the last three years (i.e., 2019 to 2021) will be addressed. These articles were stratified according to the research method such that the focus of this phase of the systematic review is to understand the new avenues of research and/or where the flow of interest from researchers on portfolio optimization is.

Among the works, there are those that talk about the use of different mechanisms, algorithms, or techniques that focus on the weight of assets or allocations. This variable weight usually makes a big difference in the results.

Another line of research that has appeared a lot is the use of machine learning or deep learning algorithms in conjunction with the mean-variance model. These machine learning algorithms are used to make predictions or classifications and, thus, choose the assets with promising results. The mean-variance model is then used to build the portfolio.

There are works that address the construction of portfolios with environmental responsibility and, therefore, restrictions regarding carbon emissions by companies are added. In the articles reviewed, the use of hybrid methods was found. Among these hybridizations, we find the line of the use of mathematical methods together with the punishment or investor sentiment.

The classifications of the approaches identified in the mentioned articles are discussed in more detail in the following subsections.

4.5.1 Machine Learning, Deep Learning, and Heuristics

Some articles discuss the use of heuristics to solve more complex models, that is, those that have more restrictions/constraints. As examples, the work of Zhai et al. [92], Gao et al. [34], Zhao et al. [95], Al Janabi et al. [5], Aguilar-Rivera and Valenzuela-Rendón [3], Khan et al. [45], Jiang et al. [42], Cui et al. [22], Chen et al. [20], Rangel-González et al. [74], Aboussalah and Lee [1], and Ünlü and Xanthopoulos [84] fall within this category.

Zhai et al. [92] use the second-order stochastic dominance rule and some real-world constraints to model the portfolio optimization problem. To solve this problem, Zhai et al. [92] use the whale optimization algorithm to optimize the portfolio and report that it is better than algorithms such as particle swarm optimization and firefly algorithm.

To address the difficulty of solving a portfolio optimization model with fuzzy theory [34, 42, 74], Gao et al. [34] propose a new mechanism-based artificial bee colony algorithm consisting of two new proposed learning strategies—driving learning and elite learning. The driving learning strategy has great potential to guide the search in promising areas whereas the elite learning strategy can gradually increase the rate of convergence without losing population diversity. The cooperation of the two approaches forms a mechanism complementing each other to improve the performance of the algorithm.

Zhao et al. [95] propose a multi-population co-evolutionary particle swarm optimization algorithm (MPCoPSO), which is based on multiple populations for multiple goals (MPMO) and has the following four advantages: (1) a binary and real hybrid coding strategy (HBR) is introduced to better represent stock selection and solution asset weight in MoCCPOP; (2) a risk-return ratio heuristic where the HBR strategy is introduced to better represent the stock selection and the weight of the solution assets in MoCCPOP; (3) a new method particle update based on bidirectional local search strategy (BLS) is designed to increase the chance of improving solution accuracy and getting closer to the global Pareto front (PF); and (4) an elite hybrid competition strategy (HEC) is proposed to help update the front, which provides more promising solutions and brings diversity to the PF.

Al Janabi et al. [5] develop a new approach to assess the liquidity-adjusted value-at-risk (LVaR) optimization of multi-asset portfolios based on vine copulas and LVaR models. This framework is applied to the stock markets of the G7 countries, gold, commodities, and Bitcoin.

Aguilar-Rivera and Valenzuela-Rendón [3] present a new algorithmic trading method based on evolutionary algorithms and portfolio theory. The model allows the inclusion of dynamic constraints such as transaction costs, portfolio imbalance, and inflation. A Monte Carlo method is proposed to deal with these types of constraints. The investment strategies method is introduced to make trading decisions based on investor preference and current market state. Preference is determined using heuristics rather than theoretical utility functions.

Quantum computing is gaining immense popularity among the scientific community as it surpasses conventional computing in efficiency and speed. All traditional computing algorithms are not directly compatible with quantum computers so it is necessary to formulate their variants using the principles of quantum mechanics. In Khan et al. [45], the search for quantum beetle antennas (QBAS) was formulated—a meta-heuristic optimization algorithm and a variant of the search for beetle antennas (BAS)—and was applied to the portfolio selection problem.

Cui et al. [22] propose a two-stage stochastic portfolio optimization model with a comprehensive set of real-world trade constraints. The model incorporates market uncertainty in terms of future asset price scenarios based on asset return distributions derived from actual market data. To solve the proposed stochastic model, a hybrid combinatorial approach was developed, which integrates a hybrid algorithm and a linear programming solver for the problem with a large number of scenarios.

In the line of machine learning, the work of Aboussalah and Lee [1] stands out through its use of a stacked deep dynamic recurrent reinforcement learning (SDDRRL) architecture to automate negotiation in real time. This reinforcement learning technique is different from traditional models in that it aims to work with continuous and multidimensional data, whereas classical techniques work with discrete data. The algorithm evaluates the market situation in real time and re-balances the portfolio while respecting the cardinality constraints. In the work of Ünlü and Xanthopoulos [84], the unsupervised learning technique of the set is used with a focus on treating the variability of the solutions through the incorporation of weights or weighting of the solutions.

4.5.2 Portfolio Optimization with Carbon Restriction

There are works that address the construction of portfolios with environmental responsibility. That is, it must be part of the company's actions that demonstrate a level of care for the environment through the preservation of ecosystems and the reduction of carbon emissions [11, 38, 50, 64].

In Monasterolo and De Angelis [64], an empirical analysis of low carbon indices is performed. The objective is to assess whether there are restrictions or measures that address the environmental factor considered in the portfolio optimization model. Thus, the financial sector should also play a more important role in de-carbonizing the global economy.

To align financial flows with a path to a low-carbon economy, investors must be able to integrate into their financial decisions additional criteria beyond return and risk to manage climate risk. Hilario-Caballero et al. [38] propose a three-criteria portfolio selection model to extend the classic Markowitz mean-variance approach to include investor preferences in portfolio carbon risk exposure as an additional criterion. To approximate the 3D Pareto front, an efficient multi-objective GA was applied. In addition, they introduce an *a posteriori* approach to incorporate investor preferences into the solution process in relation to their climate change-related preferences measured by their exposure to carbon risk and their adverse attitude to loss [38].

Bai et al. [11] present an innovative, robust portfolio allocation approach to improve the performance of the renewable energy stock portfolio in China that also considers parameter uncertainty in the optimization process. Furthermore, to make the conclusions more robust, the stock market is classified into three commonly recognized statuses: bull market, bear market, and steady market, respectively. The performance of the robust portfolio method was compared with the traditional approach of Markowitz.

4.5.3 Hybridizations: Algorithms and Strategies

In the reviewed articles, the use of hybrid methods was found [7, 39, 43, 47, 52, 59, 67, 89]. The idea of hybridizing is based on the assumption of bringing together the best in each algorithm or strategy and thus creating a robust tool that should offer better results than if they were used in isolation.

Among these hybridizations we find the line of use of mathematical methods in conjunction with punishment or investor sentiment. In Hu & Zheng [39], for example, the CAPM model is used together with the sentiment assessment. Leal et al. [47] propose a model with two-phase decision makers where one decides on unit transaction costs and the other decides on the optimal choice of assets according to returns and risk. Mendonça et al. [59] proposes an integer multi-objective average CVaR portfolio optimization model with variable cardinality constraint and rebalancing. Mendonça et al. [59] also proposes two different decision maker methods that can be used to guide and select, according to the decision maker's preferences, a solution from the non-dominated portfolios generated by an evolutionary algorithm. Decision-making methods were used to approximate investor behavior according to three functions that were chosen to represent different investor profiles (i.e., conservative, moderate, and aggressive).

Also in line with the use of mathematical techniques with expert knowledge or investor opinion, Yeh and Liu [89] employs an experimental mixing project to determine the weights of stock selection concepts, collect portfolio performance data, and build performance prediction models based on the weights of stock selection concepts. In addition, these performance prediction models and optimization techniques are employed to discover the optimal combination of stock selection concept weights that meet investor preferences.

In the study of Altinoz and Altinoz [7], two initialization approaches are proposed for multi-objective optimization algorithms to obtain better convergence and solution distribution for the portfolio optimization problem. The initial set of the population is composed of the assets with the highest income and binary combinations of assets where their sum returns the maximum income. These proposed approaches are integrated with eight different optimization algorithms and the performance of the algorithms is compared using convergence and diversity metrics.

Kara et al. [43] proposes a hybrid flow approach, in which generalized autoregressive conditional heteroskedasticity (GARCH) modeling is used to predict indicators for stocks, followed by the translation of indicator predictions to return predictions through the support vector regression. The return forecasts obtained are then fed into the Black-Litterman modeling as a vector of investor views in order to generate portfolios with continuous data. In the study of Ma et al. [52], price forecasting is combined with portfolio optimization. For the prediction, they used machine learning algorithms in conjunction with deep learning algorithms. The prediction comes in with the intention of choosing the assets that will compose the portfolio. In Paiva et al. [67], a hybrid model composed of machine learning algorithm with the mean-variance model is proposed. The machine learning algorithm makes the prediction and through the mean-variance model the optimization is done.

4.5.4 Cluster or Classification

In the clusters line of research, Duarte and De Castro [28] present a structure for asset allocation based on partial clustering algorithms. The idea is to segment assets into clusters of correlated assets, then allocate resources to, and within, each cluster. The structure allows the use of different partitional clustering algorithms as well as intra- and inter-group allocation methods. Furthermore, several evaluation criteria are considered and a specialized initialization method is proposed for the clustering algorithm.

4.5.5 Probability, Distribution and Treatment of Uncertainties

Among the articles, those dealing with problems related to uncertainty were found and therefore, terms such as "probability" and "probability distribution" appear. In this sense, the authors develop their research aimed at mitigating problems related to uncertainties and estimation errors [27, 87, 98].

Du et al. [27] propose a new method for the portfolio optimization problem with respect to distribution uncertainty. When the uncertain rate of return distribution information is observable only through a finite sample dataset, they model the portfolio selection problem with a robust optimization method from the data-driven perspective. First, they develop an ambiguous mean CVaR portfolio optimization model in which the ambiguous distribution set employed in the robust distribution model is a Wasserstein ball centered on the empirical distribution. Furthermore, the computationally tractable worst-case expectation model under the uncertainty set of a cone is derived and some theoretical conclusions of the box, budget, and ellipsoid uncertainty set are obtained.

The distribution of return on assets in the real world is generally skewed and heavy-tailed, which is far from a Gaussian distribution. Asymmetry and heavy tail are characterized by the third and fourth central moments, that is, asymmetry and kurtosis, respectively. Greater asymmetry and less kurtosis are preferable to reduce the likelihood of extreme losses. However, incorporating high-order moments into the portfolio design is very difficult due to its non-convexity and rapid computational cost increase with dimension. As a result, Zhou and Palomar [98] propose a very efficient and verifiable convergence algorithm framework based on the successive convex approximation (SCA) algorithm for solving high-order portfolios. The efficiency of the proposed algorithm framework is demonstrated by numerical experiments.

The increasing complexity and dynamic ownership of equity markets are key challenges in the financial industry, where inflexible trading strategies designed by seasoned financial professionals fail to perform satisfactorily in all market conditions. To face this challenge, adaptive stock trading strategies with deep reinforcement learning methods are proposed. For the time series nature of stock market data, the gated recurrent unit (GRU) is applied to extract informative financial characteristics, which can represent the intrinsic characteristics of the stock market for adaptive trading decisions. Furthermore, with the custom design of state and action spaces, two negotiation strategies with reinforcement learning methods are proposed as Gated Deep Q-learning negotiation strategy (GDQN) and Gated Deterministic Policy Gradient negotiation strategy (GDPG) [87].

4.5.6 Other Approaches/Strategies

There is a great deal of knowledge about the co-integration of international stock markets. In light of this concept, there are studies that provide an answer to the international portfolio optimization problem. It is likely that investors' preferences for mean, variance, asymmetry, and kurtosis will vary over time. One investor may choose to give more weight to earnings, while another may choose variation or asymmetry over return. In the study of Ashfaq et al. [8], to achieve multi-objective optimization, the study uses the goal polynomial programming model (PGP). This is the first study of its kind to consider the closing stock indices of the BRICS economies (namely Brazil, Russia, India, China, and South Africa) from January 2010 to December 2016, assigning various weights to each BRICS country under an alternative MVSK (mean, variance, asymmetry, kurtosis) structure.

Fernandez et al. [32] model investor preferences in order to find the most satisfactory portfolio from the investor's point of view when many objective functions are considered. Here, the investor's behavior towards risk, the estimates of future portfolio returns, and the risk of not achieving them are represented by means of probabilistic confidence intervals. The imperfect knowledge related to the investor's subjectivity is modeled based on range theory and the overcoming method. The proposed approach aggregates the many criteria based on the investor's particular system of preferences, producing selective pressure towards the most preferred portfolio while the investor's cognitive effort in the final selection is reduced.

Yang et al. [88] use the empirical mode decomposition (EMD) method to analyze the cross-correlation structure between different levels of fluctuations for financial assets. Correlation-based networks are then employed to determine the stock market's clustering property. Next, they propose several portfolio optimization strategies based on EMD correlation-based networks. Using the topological information of the networks, it is possible to build some portfolios with high return and low risk. In the study of Gatfaoui [36], diversification through commodities (gas and oil) and stocks is proposed to show the power of this type of diversification. One of the most important discussions in this study is that the structures of dependence on historical stock price data vary over time—this is an interesting point when thinking about developing a long-term strategy or in analyzing long-term patterns in a time series.

5 Conclusions

To provide a comprehensive systematic review with the classification of listed articles and suggestions for future research, we reviewed 73 articles published in scientific journals that address topics related to portfolio optimization: exact and heuristic methods, realistic restrictions, markets evaluated (benchmark), stock analysis types, and software/programming languages used. Although the research has not been exhaustive, it is believed that this review can facilitate the accumulation of knowledge and further studies in this area of research. The implications of this research are shown in the following paragraphs.

Regardless of the method used, the use of techniques to make estimates of future returns is a trend in studies on portfolio optimization, since the error in estimating parameters o f the model compromises the quality of optimization. Techniques such as robust optimization, Bayesian statistics, neural networks, and fuzzy logic are used to amortize the estimation error.

Themes such as robust optimization, dynamic optimization, multi-objective problems, and hybrid algorithms often appear in portfolio optimization. In addition to showing how diverse methods to solve the portfolio optimization problem are, these approaches offer a range of possibilities for combinations with each other, which can be used to increase their ability to address the problem of portfolio optimization.

The databases used to test the algorithms/tools are important for the functional analyses of the methods used. Data to test the optimization model or algorithm can be obtained on paid platforms such as Thomson Reuters or instances such as OR-Library and Fame-French that make the data available for free.

Regarding implementation, there is the possibility of using ready-made optimization software or developing custom implementations using programming languages. AMPL is a good option for the researcher who does not want to deepen his knowledge about programming and prefers to use optimization software to solve problems. Languages such as Python, R, C++, and Java require certain knowledge about programming from the user but in return, they offer advantages such as calculation speed, flexibility in model building, active communities/forums, and free availability for users.

Regarding recent works published in the last three years (2019 to 2021), we find works that talk about the use of various mechanisms, algorithms, or techniques that focus on the weight of assets or allocations. This variable weight usually makes a big difference in the results. There are also articles that use machine learning or deep learning algorithms in conjunction with the mean-variance model. These machine learning algorithms are used to make predictions or rankings and thus choose the assets with promising results and then use the mean-variance model to build the portfolio. There are also works that address the construction of portfolios with environmental responsibility and, therefore, restrictions are added regarding carbon emissions by companies. In the articles reviewed, the use of hybrid methods was found. Among these hybridizations, we find the line of use of mathematical methods in conjunction with punishment or investor sentiment.

6 Research Gaps

The gaps identified in the literature and the suggestions for future research on the theme portfolio optimization are as follows: Develop robust predictors for the risk-return binomial, aiming to increase the capacity of the model to optimize an investment portfolio; Studying the addition of constraints based on fundamental analysis of shares—that is, within this type of analysis what would be the constraints that, when added, make the mathematical model more efficient for portfolio optimization; Which constraint(s) could be added to prevent the model from seeking shares highly influenced by market volatility and, consequently, by the emotions of those people who invest; The computational complexity of the portfolio optimization problem, along with the practical importance of portfolio management, triggers researchers' motivation to design efficient algorithms. Therefore, improvements related to computational efficiency (such as parameter calibration and hybridization, among others) to deal with methodologies of algorithms, heuristic, and accuracy, are always welcome.

Although the filters and search strategies have been defined to choose the best and most current studies on the topic under discussion, it is believed that at some point, these strategies/filters may have prevented finding more relevant studies. The non-English publications filter may have deleted relevant works written in other languages. However, we are convinced that we have compiled the main work to support the discussion proposed in the systematic review protocol given in Table 1.

Acknowledgements To the Coordination for the Improvement of Higher Education Personnel (CAPES) for subsidizing this research. The authors declare that they have no conflict of interest.

References

1. Aboussalah, A.M., Lee, C.G.: Continuous control with stacked deep dynamic recurrent reinforcement learning for portfolio optimization. Expert Syst. Appl. **140**, 112891 (2020). https://doi.org/10.1016/j.eswa.2019.112891
2. Ackermann, F., Pohl, W., Schmedders, K.: Optimal and naive diversification in currency markets. Manage. Sci. **63**(10), 3347–3360 (2017). https://doi.org/10.1287/mnsc.2016.2497
3. Aguilar-Rivera, A., Valenzuela-Rendón, M.: A new multi-period investment strategies method based on evolutionary algorithms. Neural Comput. Appl. **31**(3), 923–937 (2017). https://doi.org/10.1007/s00521-017-3121-6
4. Al Janabi, M.A.: Optimal and investable portfolios: an empirical analysis with scenario optimization algorithms under crisis market prospects. Econ. Model. **40**, 369–381 (2014). https://doi.org/10.1016/j.econmod.2013.11.021
5. Al Janabi, M.A., Ferrer, R., Shahzad, S.J.H.: Liquidity-adjusted value-at-risk optimization of a multi-asset portfolio using a vine copula approach. Phys. A: Stat. Mech. Its Appl. **536**, 122579 (2019). https://doi.org/10.1016/j.physa.2019.122579
6. Algarvio, H., Lopes, F., Sousa, J., Lagarto, J.: Multi-agent electricity markets: retailer portfolio optimization using markowitz theory. Electr. Power Syst. Res. **148**, 282–294 (2017). https://doi.org/10.1016/j.epsr.2017.02.031
7. Altinoz, M., Altinoz, O.T.: Systematic initialization approaches for portfolio optimization problems. IEEE Access **7**, 57779–57794 (2019). https://doi.org/10.1109/access.2019.2914115
8. Ashfaq, S., Ayub, U., Mujtaba, G., Raza, N., Gulzar, S.: Gainers and losers with higher order portfolio risk optimization. Phys. A **563**, 125416 (2021). https://doi.org/10.1016/j.physa.2020.125416
9. Ayub, U., Shah, S.Z.A., Abbas, Q.: Robust analysis for downside risk in portfolio management for a volatile stock market. Econ. Model. **44**, 86–96 (2015). https://doi.org/10.1016/j.econmod.2014.10.001

10. Babaei, S., Sepehri, M.M., Babaei, E.: Multi-objective portfolio optimization considering the dependence structure of asset returns. Eur. J. Oper. Res. **244**(2), 525–539 (2015). https://doi.org/10.1016/j.ejor.2015.01.025
11. Bai, L., Liu, Y., Wang, Q., Chen, C.: Improving portfolio performance of renewable energy stocks using robust portfolio approach: evidence from china. Phys. A **533**, 122059 (2019). https://doi.org/10.1016/j.physa.2019.122059
12. Ban, G.Y., Karoui, N.E., Lim, A.E.B.: Machine learning and portfolio optimization. Manage. Sci. **64**(3), 1136–1154 (2018). https://doi.org/10.1287/mnsc.2016.2644
13. Bastos, L.d., Mendes, M.L., Nunes, D.R., Melo, A.C.S., Carneiro, M.P.: A systematic literature review on the joint replenishment problem solutions: 2006-2015. Production **27**(0) (2017). https://doi.org/10.1590/0103-6513.222916
14. Behr, P., Guettler, A., Miebs, F.: On portfolio optimization: imposing the right constraints. J. Bank. Financ. **37**(4), 1232–1242 (2013). https://doi.org/10.1016/j.jbankfin.2012.11.020
15. Benati, S.: Using medians in portfolio optimization. J. Oper. Res. Soc. **66**(5), 720–731 (2015). https://doi.org/10.1057/jors.2014.57
16. BenSaïda, A., Boubaker, S., Nguyen, D.K.: The shifting dependence dynamics between the g7 stock markets. Quant. Financ. **18**(5), 801–812 (2018). https://doi.org/10.1080/14697688.2017.1419628
17. Berutich, J.M., López, F., Luna, F., Quintana, D.: Robust technical trading strategies using GP for algorithmic portfolio selection. Expert Syst. Appl. **46**, 307–315 (2016). https://doi.org/10.1016/j.eswa.2015.10.040
18. Brodie, J., Daubechies, I., Mol, C.D., Giannone, D., Loris, I.: Sparse and stable markowitz portfolios. Proc. Natl. Acad. Sci. **106**(30), 12267–12272 (2009). https://doi.org/10.1073/pnas.0904287106
19. Ceren, T.Ş, Köksalan, M.: Effects of multiple criteria on portfolio optimization. Int. J. Inf. Technol. Decis. Mak. **13**(01), 77–99 (2014). https://doi.org/10.1142/s0219622014500047
20. Chen, C., Liu, D., Xian, L., Pan, L., Wang, L., Yang, M., Quan, L.: Best-case scenario robust portfolio for energy stock market. Energy **213**, 118664 (2020). https://doi.org/10.1016/j.energy.2020.118664
21. Chen, C., Zhou, Y.S.: Robust multiobjective portfolio with higher moments. Expert Syst. Appl. **100**, 165–181 (2018). https://doi.org/10.1016/j.eswa.2018.02.004
22. Cui, T., Bai, R., Ding, S., Parkes, A.J., Qu, R., He, F., Li, J.: A hybrid combinatorial approach to a two-stage stochastic portfolio optimization model with uncertain asset prices. Soft. Comput. **24**(4), 2809–2831 (2019). https://doi.org/10.1007/s00500-019-04517-y
23. Dai, Z., Wen, F.: Some improved sparse and stable portfolio optimization problems. Financ. Res. Lett. **27**, 46–52 (2018). https://doi.org/10.1016/j.frl.2018.02.026
24. DeMiguel, V., Garlappi, L., Nogales, F.J., Uppal, R.: A generalized approach to portfolio optimization: improving performance by constraining portfolio norms. Manage. Sci. **55**(5), 798–812 (2009). https://doi.org/10.1287/mnsc.1080.0986
25. DeMiguel, V., Garlappi, L., Uppal, R.: Optimal versus naive diversification: how inefficient is the 1/NPortfolio strategy? Rev. Financ. Stud. **22**(5), 1915–1953 (2007). https://doi.org/10.1093/rfs/hhm075
26. Dresch, A., Lacerda, D.P., Antunes, J.A.V.: Systematic literature review. Design Science Research, pp. 129–158. Springer International Publishing (2014). https://doi.org/10.1007/978-3-319-07374-3_7
27. Du, N., Liu, Y., Liu, Y.: A new data-driven distributionally robust portfolio optimization method based on wasserstein ambiguity set. IEEE Access **9**, 3174–3194 (2021). https://doi.org/10.1109/access.2020.3047967
28. Duarte, F.G., Castro, L.N.D.: A framework to perform asset allocation based on partitional clustering. IEEE Access **8**, 110775–110788 (2020). https://doi.org/10.1109/access.2020.3001944
29. Ertenlice, O., Kalayci, C.B.: A survey of swarm intelligence for portfolio optimization: algorithms and applications. Swarm Evol. Comput. **39**, 36–52 (2018). https://doi.org/10.1016/j.swevo.2018.01.009

30. Fabozzi, F.J., Kolm, P.N., Pachamanova, D.A., Focardi, S.M.: Robust portfolio optimization. J. Portf. Manag. **33**(3), 40–48 (2007). https://doi.org/10.3905/jpm.2007.684751
31. Fan, J., Zhang, J., Yu, K.: Vast portfolio selection with gross-exposure constraints. J. Am. Stat. Assoc. **107**(498), 592–606 (2012). https://doi.org/10.1080/01621459.2012.682825
32. Fernandez, E., Navarro, J., Solares, E., Coello Coello, C.: A novel approach to select the best portfolio considering the preferences of the decision maker. Swarm Evol. Comput. **46**, 140–153 (2019). https://doi.org/10.1016/j.swevo.2019.02.002
33. Furlan, P.K., Laurindo, F.J.B.: Agrupamentos epistemológicos de artigos publicados sobre big data analytics. Transinformação **29**(1), 91–100 (2017). https://doi.org/10.1590/2318-08892017000100009
34. Gao, W., Sheng, H., Wang, J., Wang, S.: Artificial bee colony algorithm based on novel mechanism for fuzzy portfolio selection. IEEE Trans. Fuzzy Syst. **27**(5), 966–978 (2019). https://doi.org/10.1109/tfuzz.2018.2856120
35. García, F., Guijarro, F., Oliver, J.: Index tracking optimization with cardinality constraint: a performance comparison of genetic algorithms and tabu search heuristics. Neural Comput. Appl. **30**(8), 2625–2641 (2017). https://doi.org/10.1007/s00521-017-2882-2
36. Gatfaoui, H.: Diversifying portfolios of U.S. stocks with crude oil and natural gas: a regime-dependent optimization with several risk measures. Energy Econ. **80**, 132–152 (2019). https://doi.org/10.1016/j.eneco.2018.12.013
37. Goodman, D.: Brian: a simulator for spiking neural networks in python. Front. Neuroinform. **2** (2008). https://doi.org/10.3389/neuro.11.005.2008
38. Hilario-Caballero, A., Garcia-Bernabeu, A., Salcedo, J.V., Vercher, M.: Tri-criterion model for constructing low-carbon mutual fund portfolios: a preference-based multi-objective genetic algorithm approach. Int. J. Environ. Res. Public Health **17**(17), 6324 (2020). https://doi.org/10.3390/ijerph17176324
39. Hu, W., Zheng, Z.: Expectile CAPM. Econ. Model. **88**, 386–397 (2020). https://doi.org/10.1016/j.econmod.2019.09.049
40. Hu, Y., Liu, K., Zhang, X., Su, L., Ngai, E., Liu, M.: Application of evolutionary computation for rule discovery in stock algorithmic trading: a literature review. Appl. Soft Comput. **36**, 534–551 (2015). https://doi.org/10.1016/j.asoc.2015.07.008
41. Ji, R., Lejeune, M.A., Prasad, S.Y.: Properties, formulations, and algorithms for portfolio optimization using mean-gini criteria. Ann. Oper. Res. **248**(1–2), 305–343 (2016). https://doi.org/10.1007/s10479-016-2230-4
42. Jiang, M., Liu, W., Xu, W., Chen, W.: Improved multiobjective bat algorithm for the credibilistic multiperiod mean-VaR portfolio optimization problem. Soft. Comput. **25**(8), 6445–6467 (2021). https://doi.org/10.1007/s00500-021-05638-z
43. Kara, M., Ulucan, A., Atici, K.B.: A hybrid approach for generating investor views in black–litterman model. Expert Syst. Appl. **128**, 256–270 (2019). https://doi.org/10.1016/j.eswa.2019.03.041
44. Karakalidis, A., Sifaleras, A.: Solving portfolio optimization problems using AMPL. Operational Research in Business and Economics, pp. 167–184. Springer International Publishing (2016). https://doi.org/10.1007/978-3-319-33003-7_8
45. Khan, A.T., Cao, X., Li, S., Hu, B., Katsikis, V.N.: Quantum beetle antennae search: a novel technique for the constrained portfolio optimization problem. Sci. China Inf. Sci. **64**(5) (2021). https://doi.org/10.1007/s11432-020-2894-9
46. Kumar, D., Mishra, K.: Portfolio optimization using novel co-variance guided artificial bee colony algorithm. Swarm Evol. Comput. **33**, 119–130 (2017). https://doi.org/10.1016/j.swevo.2016.11.003
47. Leal, M., Ponce, D., Puerto, J.: Portfolio problems with two levels decision-makers: optimal portfolio selection with pricing decisions on transaction costs. Eur. J. Oper. Res. **284**(2), 712–727 (2020). https://doi.org/10.1016/j.ejor.2019.12.039
48. Levy, M., Kaplanski, G.: Portfolio selection in a two-regime world. Eur. J. Oper. Res. **242**(2), 514–524 (2015). https://doi.org/10.1016/j.ejor.2014.10.012

49. Li, Q., Bao, L.: Enhanced index tracking with multiple time-scale analysis. Econ. Model. **39**, 282–292 (2014). https://doi.org/10.1016/j.econmod.2014.03.009
50. Liagkouras, K., Metaxiotis, K., Tsihrintzis, G.: Incorporating environmental and social considerations into the portfolio optimization process. Ann. Oper. Res. (2020). https://doi.org/10.1007/s10479-020-03554-3
51. Liu, Y.J., Zhang, W.G.: A multi-period fuzzy portfolio optimization model with minimum transaction lots. Eur. J. Oper. Res. **242**(3), 933–941 (2015). https://doi.org/10.1016/j.ejor.2014.10.061
52. Ma, Y., Han, R., Wang, W.: Portfolio optimization with return prediction using deep learning and machine learning. Expert Syst. Appl. **165**, 113973 (2021). https://doi.org/10.1016/j.eswa.2020.113973
53. Macedo, L.L., Godinho, P., Alves, M.J.: Mean-semivariance portfolio optimization with multiobjective evolutionary algorithms and technical analysis rules. Expert Syst. Appl. **79**, 33–43 (2017). https://doi.org/10.1016/j.eswa.2017.02.033
54. Mansini, R., Ogryczak, W., Speranza, M.G.: Twenty years of linear programming based portfolio optimization. Eur. J. Oper. Res. **234**(2), 518–535 (2014). https://doi.org/10.1016/j.ejor.2013.08.035
55. Markowitz, H.: Portfolio selection. J. Financ. **7**(1), 77–91 (1952). https://doi.org/10.1111/j.1540-6261.1952.tb01525.x
56. Marzban, S., Mahootchi, M., Khamseh, A.A.: Developing a multi-period robust optimization model considering american style options. Ann. Oper. Res. **233**(1), 305–320 (2013). https://doi.org/10.1007/s10479-013-1461-x
57. McKinney, W.: Data structures for statistical computing in python. In: Proceedings of the 9th Python in Science Conference, vol. 445, pp. 51–56 (2010)
58. Meghwani, S.S., Thakur, M.: Multi-objective heuristic algorithms for practical portfolio optimization and rebalancing with transaction cost. Appl. Soft Comput. **67**, 865–894 (2018). https://doi.org/10.1016/j.asoc.2017.09.025
59. Mendonça, G.H., Ferreira, F.G., Cardoso, R.T., Martins, F.V.: Multi-attribute decision making applied to financial portfolio optimization problem. Expert Syst. Appl. **158**, 113527 (2020). https://doi.org/10.1016/j.eswa.2020.113527
60. Merton, R.C.: Optimum consumption and portfolio rules in a continuous-time model. Stochastic Optimization Models in Finance, pp. 621–661. Elsevier, Amsterdam (1975). https://doi.org/10.1016/b978-0-12-780850-5.50052-6
61. Milhomem, D.A., Dantas, M.J.P.: Analysis of new approaches used in portfolio optimization: a systematic literature review. Production **30** (2020). https://doi.org/10.1590/0103-6513.20190144
62. Mishra, S.K., Panda, G., Majhi, B.: Prediction based mean-variance model for constrained portfolio assets selection using multiobjective evolutionary algorithms. Swarm Evol. Comput. **28**, 117–130 (2016). https://doi.org/10.1016/j.swevo.2016.01.007
63. Mitchell, S., O'Sullivan, M., Dunning, I.: PuLP: a linear programming toolkit for python. Technical report, The University of Auckland (2011)
64. Monasterolo, I., de Angelis, L.: Blind to carbon risk? an analysis of stock market reaction to the Paris Agreement. Ecol. Econ. **170**, 106571 (2020). https://doi.org/10.1016/j.ecolecon.2019.106571
65. Özden Gür, A., Akçay, Y., Sayman, S., Yılmaz, E., Özçelik, M.H.: Cross-selling investment products with a win-win perspective in portfolio optimization. Oper. Res. **65**(1), 55–74 (2017). https://doi.org/10.1287/opre.2016.1556
66. Pai, G.A.V.: Fuzzy decision theory based metaheuristic portfolio optimization and active rebalancing using interval type-2 fuzzy sets. IEEE Trans. Fuzzy Syst. **25**(2), 377–391 (2017). https://doi.org/10.1109/tfuzz.2016.2633972
67. Paiva, F.D., Cardoso, R.T.N., Hanaoka, G.P., Duarte, W.M.: Decision-making for financial trading: a fusion approach of machine learning and portfolio selection. Expert Syst. Appl. **115**, 635–655 (2019). https://doi.org/10.1016/j.eswa.2018.08.003

68. Pedregosa, F., Varoquaux, G., Gramfort, A., Michel, V., Thirion, B., Grisel, O., Blondel, M., Müller, A., Nothman, J., Louppe, G., Prettenhofer, P., Weiss, R., Dubourg, V., Vanderplas, J., Passos, A., Cournapeau, D., Brucher, M., Perrot, M., Duchesnay, E.: Scikit-learn: machine learning in python. J. Mach. Learn. Res. (2011)
69. Pekár, J., Čičková, Z., Brezina, I.: Portfolio performance measurement using differential evolution. CEJOR **24**(2), 421–433 (2015). https://doi.org/10.1007/s10100-015-0393-8
70. Pflug, G., Wozabal, D.: Ambiguity in portfolio selection. Quant. Financ. **7**(4), 435–442 (2007). https://doi.org/10.1080/14697680701455410
71. Pflug, G.C., Pichler, A., Wozabal, D.: The 1/N investment strategy is optimal under high model ambiguity. J. Bank. Financ. **36**(2), 410–417 (2012). https://doi.org/10.1016/j.jbankfin.2011.07.018
72. Pouya, A.R., Solimanpur, M., Rezaee, M.J.: Solving multi-objective portfolio optimization problem using invasive weed optimization. Swarm Evol. Comput. **28**, 42–57 (2016). https://doi.org/10.1016/j.swevo.2016.01.001
73. Qu, B.Y., Zhou, Q., Xiao, J.M., Liang, J.J., Suganthan, P.N.: Large-scale portfolio optimization using multiobjective evolutionary algorithms and preselection methods. Math. Probl. Eng. **2017**, 1–14 (2017). https://doi.org/10.1155/2017/4197914
74. Rangel-González, J.A., Fraire, H., Solís, J.F., Cruz-Reyes, L., Gomez-Santillan, C., Rangel-Valdez, N., Carpio-Valadez, J.M.: Fuzzy multi-objective particle swarm optimization solving the three-objective portfolio optimization problem. Int. J. Fuzzy Syst. **22**(8), 2760–2768 (2020). https://doi.org/10.1007/s40815-020-00928-4
75. Ren, F., Lu, Y.N., Li, S.P., Jiang, X.F., Zhong, L.X., Qiu, T.: Dynamic portfolio strategy using clustering approach. PLoS ONE **12**(1), e0169299 (2017). https://doi.org/10.1371/journal.pone.0169299
76. Reveiz-Herault, A.: An active asset management investment process for drawdown-averse investors. Intell. Syst. Account., Financ. Manag. **23**(1–2), 85–96 (2015). https://doi.org/10.1002/isaf.1375
77. Rubio, A., Bermúdez, J.D., Vercher, E.: Forecasting portfolio returns using weighted fuzzy time series methods. Int. J. Approx. Reason. **75**, 1–12 (2016). https://doi.org/10.1016/j.ijar.2016.03.007
78. Rubio, A., Bermúdez, J.D., Vercher, E.: Improving stock index forecasts by using a new weighted fuzzy-trend time series method. Expert Syst. Appl. **76**, 12–20 (2017). https://doi.org/10.1016/j.eswa.2017.01.049
79. Saborido, R., Ruiz, A.B., Bermúdez, J.D., Vercher, E., Luque, M.: Evolutionary multi-objective optimization algorithms for fuzzy portfolio selection. Appl. Soft Comput. **39**, 48–63 (2016). https://doi.org/10.1016/j.asoc.2015.11.005
80. Sharma, C., Banerjee, K.: A study of correlations in the stock market. Phys. A **432**, 321–330 (2015). https://doi.org/10.1016/j.physa.2015.03.061
81. Silva, A., Neves, R., Horta, N.: A hybrid approach to portfolio composition based on fundamental and technical indicators. Expert Syst. Appl. **42**(4), 2036–2048 (2015). https://doi.org/10.1016/j.eswa.2014.09.050
82. Sun, X., Liu, Z.: Optimal portfolio strategy with cross-correlation matrix composed by DCCA coefficients: evidence from the chinese stock market. Phys. A **444**, 667–679 (2016). https://doi.org/10.1016/j.physa.2015.10.065
83. Thi, H.A.L., Moeini, M.: Long-short portfolio optimization under cardinality constraints by difference of convex functions algorithm. J. Optim. Theory Appl. **161**(1), 199–224 (2012). https://doi.org/10.1007/s10957-012-0197-0
84. Ünlü, R., Xanthopoulos, P.: A reduced variance unsupervised ensemble learning algorithm based on modern portfolio theory. Expert Syst. Appl. **180**, 115085 (2021). https://doi.org/10.1016/j.eswa.2021.115085
85. Uryasev, S.: Conditional value-at-risk: optimization algorithms and applications. In: Proceedings of the IEEE/IAFE/INFORMS 2000 Conference on Computational Intelligence for Financial Engineering (CIFEr) (Cat. No.00TH8520). IEEE (2000). https://doi.org/10.1109/cifer.2000.844598

86. Vercher, E., Bermúdez, J.D.: Portfolio optimization using a credibility mean-absolute semi-deviation model. Expert Syst. Appl. **42**(20), 7121–7131 (2015). https://doi.org/10.1016/j.eswa.2015.05.020
87. Wu, X., Chen, H., Wang, J., Troiano, L., Loia, V., Fujita, H.: Adaptive stock trading strategies with deep reinforcement learning methods. Inf. Sci. **538**, 142–158 (2020). https://doi.org/10.1016/j.ins.2020.05.066
88. Yang, L., Zhao, L., Wang, C.: Portfolio optimization based on empirical mode decomposition. Phys. A **531**, 121813 (2019). https://doi.org/10.1016/j.physa.2019.121813
89. Yeh, I.C., Liu, Y.C.: Discovering optimal weights in weighted-scoring stock-picking models: a mixture design approach. Financ. Innov. **6**(1) (2020). https://doi.org/10.1186/s40854-020-00209-x
90. Yu, D., Wang, W., Zhang, W., Zhang, S.: A bibliometric analysis of research on multiple criteria decision making. Curr. Sci. **114**(04), 747 (2018). https://doi.org/10.18520/cs/v114/i04/747-758
91. Zadeh, L.: Fuzzy sets. Inf. Control **8**(3), 338–353 (1965). https://doi.org/10.1016/s0019-9958(65)90241-x
92. Zhai, Q.H., Ye, T., Huang, M.X., Feng, S.L., Li, H.: Whale optimization algorithm for multi-constraint second-order stochastic dominance portfolio optimization. Comput. Intell. Neurosci. **2020**, 1–19 (2020). https://doi.org/10.1155/2020/8834162
93. Zhang, W.G., Liu, Y.J.: Credibilitic mean-variance model for multi-period portfolio selection problem with risk control. OR Spectr. **36**(1), 113–132 (2013). https://doi.org/10.1007/s00291-013-0335-6
94. Zhang, Y., Li, X., Guo, S.: Portfolio selection problems with markowitz's mean–variance framework: a review of literature. Fuzzy Optim. Decis. Making **17**(2), 125–158 (2017). https://doi.org/10.1007/s10700-017-9266-z
95. Zhao, H., Chen, Z.G., Zhan, Z.H., Kwong, S., Zhang, J.: Multiple populations co-evolutionary particle swarm optimization for multi-objective cardinality constrained portfolio optimization problem. Neurocomputing **430**, 58–70 (2021). https://doi.org/10.1016/j.neucom.2020.12.022
96. Zhao, L., Li, W., Fenu, A., Podobnik, B., Wang, Y., Stanley, H.E.: The q-dependent detrended cross-correlation analysis of stock market. J. Stat. Mech: Theory Exp. **2018**(2), 023402 (2018). https://doi.org/10.1088/1742-5468/aa9db0
97. Zhao, L., Wang, G.J., Wang, M., Bao, W., Li, W., Stanley, H.E.: Stock market as temporal network. Phys. A **506**, 1104–1112 (2018). https://doi.org/10.1016/j.physa.2018.05.039
98. Zhou, R., Palomar, D.P.: Solving high-order portfolios via successive convex approximation algorithms. IEEE Trans. Signal Process. **69**, 892–904 (2021). https://doi.org/10.1109/tsp.2021.3051369

A Temporal Knapsack Approach to Defence Portfolio Selection

Gregory Calbert, Terence Weir, Ivan L. Garanovich, and Carlos C. N. Kuhn

Abstract In this chapter we describe the origins and mathematics used in the development of the New Investments to Risked Options (NITRO) portfolio selection tool. This tool was used by the Australian Department of Defence in 2019 for a major review of Australia's military force structure called the Force Structure Plan 2020. This is a strategic decision making process that specifies the requirements for the future force structure and identifies, selects, and schedules projects to satisfy those requirements. In order to satisfy organisational needs, the tool was implemented in Microsoft Excel® using the What's*Best!*® optimisation add-in. In this chapter we provide background on the portfolio optimisation problem in Defence. We discuss the challenges of implementing a project portfolio selection and scheduling tool in the Excel® environment along with simple approaches to the visualisation of optimised portfolios. We examine the performance differences between this tool and other commercial solvers. We also analyse the trade-off of using relaxed versus integer decision variables and the use of budget slack on the portfolio selection. We conclude by discussing current portfolio optimisation issues and challenges, along with the viability of the Excel® environment for Defence portfolio selection.

G. Calbert (✉) · T. Weir · I. L. Garanovich · C. C. N. Kuhn
Joint Operations and Analysis Division, Defence Science and Technology Group, Department of Defence, Canberra, Australia
e-mail: greg.calbert@dst.defence.gov.au

T. Weir
e-mail: terence.weir@dst.defence.gov.au

I. L. Garanovich
e-mail: ivan.garanovich@dst.defence.gov.au

C. C. N. Kuhn
e-mail: carlos.kuhn@dst.defence.gov.au

K. R. Harrison et al., *Evolutionary and Memetic Computing for Project Portfolio Selection and Scheduling*, Adaptation, Learning, and Optimization 26,
https://doi.org/10.1007/978-3-030-88315-7_7

1 Introduction

In any large company or organisation the question of which activities or projects best suit overall objectives is a complex and challenging task. The challenges in choosing a subset from a collection of investments or projects are many. Organisations will have a finite resource base, in terms of staff, and dollars. Activities or projects will inherit a cost and staff profile spanning timescales of days, months, or years. In an uncertain world, the benefit of project resource investments will invariably be uncertain. Furthermore, there will be temporal or other structural dependencies between projects if activities are interlinked.

In the parlance of management science, such a collection of projects or investments is often termed a *portfolio* [50]. Business managers typically seek to optimise a portfolio, which means choosing a subset from a set of possible investments, subject to budgetary bounds, that maximises the financial return over a specified timescale. There is a substantial literature on financial portfolio investment starting with the seminal work of Harry Markowitz in 1952 [32]. Applications of portfolio optimisation are found commonly in financial markets analysis [42], the nuclear power industry [27], research and development [23] and increasingly in Defence [21], but techniques of portfolio optimisation can also be found in disparate areas such as cryptography [38] and quantum computing [35].

To formulate a real-world problem in a mathematical model, some authors like to break it into four tasks [58]:

- recognition of the problem;
- formulation of a mathematical model;
- solution of the mathematical problem; and
- translation of the results back into the context of the original problem.

Of course, this is not the only way to attack a problem, but it summarises well enough the applied mathematician's approach. Investment selection problems with budgetary constraints, in a large variety of cases, can be formulated as a set of linear equations and inequalities which can be solved by what is known as the linear programming formulation. Indeed, one of the earliest and *outstanding* [14] applications of linear programming was to investment planning within Defence organisations [47].

In the context of Defence applications, a portfolio has a similar albeit modified definition. Though the objective is to choose from a subset of possible investments or projects, each with an associated temporal cost, there is no fiscal *return on investment*, rather there is a *value*, either static or dynamic associated with each project. Such values can be elicited through war-gaming and the use of other analysis methods such as Bayesian networks or through the opinions of subject matter experts (SMEs). A thorough review of contemporary Defence applications is in [21].

This portfolio selection problem is an application of a well known combinatorial optimisation problem known as the knapsack problem [26]. At its root is the problem of selecting from a set of items, each with a given weight and value, that maximises the total value while not exceeding the specified weight limit of the knapsack. In the case of Defence applications the investment projects are the "items" , each with an assigned value and cost, and the available budget is the "weight limit".

Knapsack problems have been extensively studied, especially with increasing computer power, in the last few decades, from both a theoretical and practical viewpoint. The knapsack problem's deceptive simplicity but inherent complexity invites investigation of its combinatorial features. Knapsack problems arise commonly in capital budgeting, electricity generation, and transportation as well as many other industrial applications. Comprehensive texts on knapsack problems are [26, 33]. There are a number of named extensions to the knapsack problem explored in [26, 33] including the Multiple Knapsack Problem in which there are multiple knapsacks to be filled and the Generalised Assignment Problem, which considers items having vector values and weights. Both of these latter problems are closely related to the practical problem considered in this chapter.

The knapsack problem is known to be *NP*-hard. Consequently, an arbitrary instance cannot be solved in a time bounded by a polynomial, unless $P = NP$. However, it admits a pseudo-polynomial algorithm, i.e., an algorithm whose time (and space) complexity is bounded by a polynomial on the value of the input, rather than the length of the input [33]. References [26, 33] state a number of algorithms developed to solve knapsack problems and their variants. Popular practical approaches to solve these (either exactly or approximately) include a range of specifically designed methods such as Dynamic Programming and Branch and Bound. Heuristic approaches such as memetic algorithms have also been studied and provide useful solution strategies. Linear programming relaxation can find approximate solutions. A survey of techniques can be found at [63].

Several factors make Defence investment selection problems challenging when compared to, say, portfolio selection for financial investments. The planning for Defence acquisitions is a complex process that must satisfy numerous objectives and involves a diverse and large range of stakeholders. Generally Defence deals with large numbers of simultaneously occurring projects; often it needs to also consider alternatives of the same project as a potential investment, but of course only one alternative can be selected. Measuring the value of Defence projects is difficult because, generally, projects need to be selected to satisfy multiple objectives, while taking account of dependencies, synergies and precedence constraints with other projects, Defence force structure assets, infrastructure, and workforce. A single measure of value is often inappropriate. The costs of Defence projects can be difficult to accurately assess because sometimes technologies have not been fully developed at the time of planning and costs are often non-linear and non-additive. The ability to reasonably value large and small investments in a common comparable scale across different domains also raises challenges - for example, how does one determine the value of a new frigate compared to that of new land-based weapons. The portfolio selection process covers long time-frames in which the budget will not necessarily be completely known. How projects will perform (i.e., achieve their target benefits [66]) and the growth in key capability factors such as technology change is also uncertain. From a budget management and cash flow perspective, year-to-year expenditure shortfalls can not generally be recovered. A good explanation of these and other relevant issues is given in [21, 49].

Following this introduction, in Sect. 2, we provide background to the portfolio selection challenges in the Australian Department of Defence (DoD). We then describe the mathematical formulation of the optimisation problem in Sect. 3 and the practicalities of solving the binary optimisation problem are discussed in Sect. 4. The performance of algorithms on the optimisation problem are evaluated in Sect. 5. A discussion on future work is summarised in Sect. 6.

2 Project and Portfolio Selection in DoD

A general framework for project and portfolio selection has been developed by [2]. The framework organises project selection techniques logically into a flexible process to support the project portfolio selection. There are five major process stages in the framework. The first three stages focus on potential project identification through *pre-screening, individual project analysis*, and *screening*. These activities are generally performed off-line and can be conducted by analysts supporting decision makers. Pre-screening applies strategy guidance to ensure that any project being considered fits the strategic focus of the portfolio, has undergone a preliminary analysis for risk and cost, and has organisational support through a project champion or strategic leader. At the Individual Project Analysis stage, projects are scored on common criteria to unify comparison using concepts such as rate of return, net present value, and utility to determine a comparable value. This stage can be resource intensive if project analysis require inputs from many sections of the organisation. Finally, during the Screening stage, projects may be eliminated if they do not meet set standards or required characteristics. The last two stages *optimal portfolio selection* and *portfolio adjustment* use the identified and screened projects to create the portfolio. With suitable decision support tools these processes can be performed in group settings of decision makers. Some form of optimisation model can be used to select projects according to their value scored from earlier stages while taking account of available organisational resources, required timings, and project precedence and constraints. The final stage of Portfolio Adjustment allows decision makers to amend the portfolio based on non-quantified or other factors such as specialised knowledge or experience. Amendments can be the addition or the deletion of project or other adjustments such as changes to timings or resources. A survey of project portfolio selection and optimisation studies focusing on the evaluation criteria, selection approach, solution approach, uncertainty modelling, and applications is presented at [39].

The Australian Department of Defence (DoD) Strategy Framework 2017 [54] outlines how Government guidance and direction on strategic policy, including capability planning, is delivered across Defence. The Strategy Framework highlights the key documents guiding the Government's strategic Defence policy. Figure 1 illustrates the key documents and information flows within Defence"s planning framework.

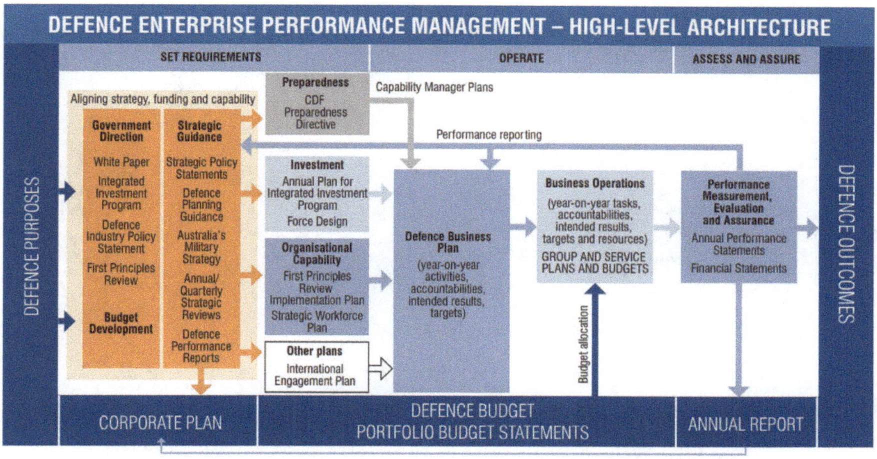

Fig. 1 DoD, The Strategy Framework 2017, p. 4

A central part of the Strategy Framework in Defence is investment planning which is generated by its force design function and promulgated in its resulting investment plan. The DoD follows a strategic process to design and implement its portfolio of projects [67]. Known as the Defence Capability Assessment Program (DCAP) [24], the force design process translates Government and Defence policy guidance into a vision of the future force structure that can be articulated to Government. The output of the force design process is expressed in the Integrated Investment Program (IIP), which describes DoD's capital investment plan in major equipment, information and communications technology, facilities, and workforce. It also programs sustainment funding for in-service and planned investments. The IIP forms the basis for the future force structure [51]. Updates to the IIP are presented to Government twice a year for approval in line with the budget cycle.

Force design follows an iterative process of risk assessment and prioritisation, options development, and portfolio design. The portfolio selection process essentially consists of three phases of interlinked activities:

1. strategic considerations that present decision makers with the security environment in which the DoD operates, its challenges, risks, and opportunities,
2. individual project evaluation that quantifies how risks are mitigated, how opportunities may be captured, and how projects contribute to organisational goals, and
3. portfolio selection in which projects are selected to best satisfy organisational requirements while observing resource constraints.

The first phase identifies strategic requirements, risks, opportunities, and overall budget allocation for the portfolio. The second phase uses techniques to identify and evaluate potential projects either individually or as collections of projects that may display synergies. This phase also identifies projects in the current portfolio

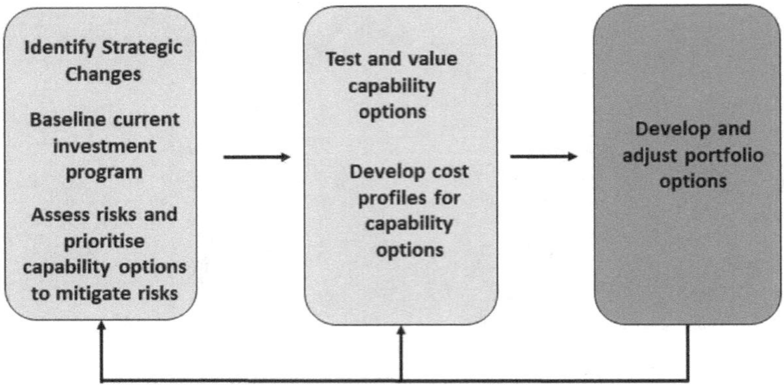

Fig. 2 This figure summarises the force design process for portfolio development and adjustment and forms the basis for the DCAP

considered for removal or change to timing or funding. In the third phase portfolios are created based on candidate project parameters; including interactions with other projects through interdependencies in time or function and resource constraints such as the budget available or workforce required to implement the projects. The model at Fig. 2 summarises the relevant aspects of the broader model in [2] for the DoD portfolio construction and adjustment needs. Considerable guidance is given in [2] regarding the design of activities and support for decision makers in each phase. Guidance on the design of a suitable decision support system for portfolio construction is given at [1].

The DCAP consists of a number of interdependent steps that might vary depending on the portfolio design requirements [10]. Each activity requires stakeholder input and concludes with decisions taken by senior committees. Analysis is focused on a small number of key points of time in the future, known as epochs.

- Identify Change. Inputs to the process such as changes to strategic direction, supporting Concepts, updates to the Defence Workforce Plan, and developments in industry capability are considered in the context of the current and future force.
- Qualify Risk. Risks to the force are qualified by mapping them through Joint Capability Effects.
- Prioritise Risk. Risks are prioritised in warfighting domains.
- Develop Options. A range of options for capability investment, divestment, and offsets are developed. Options development is based on adjusting resourcing levels or amending in-service or planned capabilities or adding new capabilities. Cost estimates are included as part of this step.
- Test Options. A baseline assessment of the future force is conducted to confirm the severity of risks. Costed options from previous steps are assessed against risks. The analysis is conducted by SMEs using tools such as facilitated decision conferences and war gaming.

- Identify Offsets. Offsets created from the current and future force are identified to ensure that the new or changed investments can be accommodated within the existing budget profile.
- Test Portfolio Options. Testing is conducted to identify the net effect of proposed new investments against proposed offsets.
- Decision. A selected set of options to address prioritised risk is presented to senior committees for consideration and to recommend options for Government decision.

The depth of the analysis in each step of the DCAP varies over a cycle of several years to satisfy strategic change assessments and to accord with organisational resource constraints. Activities such as Force Structure Plan 2020 (FSP2020), are intense complete cycles of the DCAP. FSP2020 was a designated campaign to review and update the IIP over a twenty year time horizon, using force design concepts based on Capabilities Based Planning (CBP) [57] and revised organisation arrangements formalised in the First Principles Review [37]. The FSP2020 activity included updated Government strategic guidance for DoD and was informed by assessments and analysis from Australia's security agencies and partners. SME participated in structured elicitation sessions [62] and analysts collected a comprehensive body of evidence for assessing the current and proposed force structures [41]. This approach facilitated whole of force analysis allowing the conclusions drawn from experiments to aid strategic decision making. The FSP2020 tested both the current force and proposed new capabilities to be acquired in the future against a broad range of classified scenarios over a 20 year period.

In the DoD, projects are vehicles to deliver enhanced Defence force capability. In the Defence context, capability has a rather narrow definition: "the power to achieve a desired operational effect in a nominated environment, within a specified time, and to sustain that effect for a designated period" [56]. Capability is typically managed and assessed with regard to a number of dimensions or integrative elements. In DoD capability is generated by combining nine elements or resource types known as Fundamental Inputs to Capability (FIC) [10]:

- personnel, for the delivery, operation, sustainment, and disposal of the capability;
- organisation, for employment of the capability within an organisational structure that is flexible enough to meet the endorsed capability requirements stemming from the original need;
- collective training, for supporting the capability by a defined collective training regime to a validated performance level against the Defence planning requirements and based on the original need;
- major systems, for enabling effective generation and employment of the capability through significant platforms, fleets of equipment, and operating systems;
- supplies, for providing sufficient classes of supply to maintain the capability at designated preparedness states;
- facilities and training areas, for providing the infrastructure necessary to support the delivery, sustainment, and operation of the capability;
- support, to provide engineering support; maintenance support; logistics support; technical data and information systems support;

- command and management, including organisational, command, control, and communications arrangements at all levels to safely and effectively employ the capability, including its integration with other capabilities across Defence;
- industry, to provide adequate support for the capability including the resilience and capacity of industry, such as the reliability and health of supply chains.

Viewed as a system, capability is the effects provided by interdependent FIC.

CBP is "a framework for planning under uncertainty that provides capabilities suitable for a wide range of modern-day challenges and circumstances within an economic framework that necessitates choice" [9]. The framework is employed in many Western nations as best-practice for Defence planning [46] and has also been used in non-Defence related applications, demonstrating its widening appeal as a planning tool [55]. CBP seeks to identify dependencies between strategies, mission objectives, and high-level requirements without prematurely prescribing the use of specific resources [6]. This process facilitates the development of options for meeting strategic goals.

Scenarios are a key feature of the CBP process. Unlike "point" scenarios, CBP scenarios emphasise uncertainty rather than explicit conditions [8]. Options are generated in response to the threat context outlined in strategic guidance and are subsequently evaluated and tested in the context of a robust set of threat scenarios.

The concepts of CBP are closely related to that of Project Portfolio Management [67] in which the portfolio is cognisant of interactions between projects and risk exposure is explicitly treated. Correspondingly, Defence portfolios should address a wide range of contingencies and their construction should take advantage of the synergies that can be achieved by capabilities acting in unison.

Project identification in DoD accords with guidance provided by project management literature [70] and can occur in two ways. It may be driven by strategy [52] and performed top-down [71], or generated bottom-up in response to operational weaknesses or identified opportunities using design thinking processes [4]. Often this is done in generating service specific responses. For example, [65] argues that Royal Australian Air Force SMEs perform a critical function in reconciling top-down strategy to meet bottom-up creativity and innovation. DoD provides further support mechanisms for force design personnel to identify opportunities through the Defence Innovation Hub [22] and through exploration of emerging technologies by the Defence Science and Technology Group [36]. Following identification, projects must be evaluated as to their worth or the benefits they will provide to Defence. The evaluation takes into account a range of criteria that include risk mitigation, security enhancement, alignment with Government, and Defence goals and contribution to national economic objectives. Typical evaluation methods include Multi-Criteria Decision Analysis [5, 59]. This part of the project identification process is organisationally and workforce intensive, requiring the inputs from many SMEs and the blending of strategy, futurism, concept development, gaming, and operations and tactics [9, 41]. In DoD this typically takes place over a number of weeks or months as it requires extensive consultation.

Despite the degree of formalisation in the force design process, [67] has pointed out that project initiation lacks process assurance for how this can be done effectively and methodologically and have developed a novel process for identification, measurement, and ranking of projects. According to research sponsored by the Project Management Institute [64], this "front-end" process is critical to the strategic success of the project but its part of the project life-cycle is not well understood. This initialisation of acquisition planning is of increasing interest because of its importance to project selection and strategic success. A good description of the state of the art is in [68] that also provides an hypothetical Defence case study based on the Australian force design process. They conclude that the Defence process in [10] has important implications for organisations seeking a competitive advantage through the generation of new project ideas. Measuring the benefits of projects at the initiation stage of their life-cycle is important for valuing projects but presents challenges in the Defence context as pointed out by [48]. Defence outcomes can be fuzzy and project contributions are difficult to measure especially when objectives overlap. The typical approach to deal with this difficulty is to measure some important aspects of project outcomes, then apply sensitivity analysis and subject matter opinion to generated results to test for their credibility. Criteria definition is important in that they are, as much as possible, measurable and understandable in the context of the decision and aligned with objectives. Reference [69] offers some advice on this issue.

Within the force design process model, project meta-data, value, and cost, precedence and timing constraints, and user preferences are inputs to a mathematical optimisation model that subsequently selects projects and their timings to construct a project portfolio. The portfolio can be adjusted by users and user input is supported by a set of portfolio data and visualisations depicting views of the portfolio's parameters and characteristics.

3 Problem Formulation

Force designers in the Australian DoD worked with the Defence Science and Technology Group in 2019 to develop a portfolio optimisation application as part of the FSP2020 major review of Australia's military force structure [53]. The optimisation application was named *New Investments to Risked Options* or by the abbreviation *NITRO*. As with many organisations, Microsoft Excel® is a widely used application within Defence and NITRO was implemented in Excel. While Excel has the add-in Solver® for optimisation problems, it is restricted in problem size due to limits on the number of allowable decision variables, and we decided to use the What's*Best!*® optimisation add-in to solve the large optimisation problems. What's*Best!*® is developed to allow the user to build large scale optimisation models in a free form layout within an Excel spreadsheet, and its first version was released in 1985 by LINDO Systems, Inc. Details of how this was implemented will be described in Sect. 4. What is described below is the mathematical formulation of the force design problem.

The simplest formulation of the portfolio design problem is described as follows. Suppose one has a set of projects, labelled P. Each project i in P has an associated cost, either positive or negative (for cost savings projects) termed c_i. The value of the project calculated through simulation or elicited from SMEs is termed V_i. The overall budget of the portfolio is constrained by the value B. The binary decision variable is

$$x_i = \begin{cases} 1 & \text{if project } i \text{ is in the portfolio,} \\ 0 & \text{otherwise.} \end{cases} \tag{1}$$

In its simplest form, the portfolio design problem is just the knapsack problem in combinatorial optimisation, which seeks a subset of projects $Q \subset P$ such that

$$\max_{Q \subseteq P} \left[\sum_{Q} V_i x_i \right]$$

$$\sum_{Q} c_i x_i \leq B.$$

Force designers wanted to know not only which projects should be considered in the portfolio but also the schedule of such projects. As such, the approach considered in this chapter is similar to that of the multiple knapsack problem [26]. In the multiple knapsack problem one considers the allocation of set of items, which in this case are project timings allocated to a set of knapsacks, which are the set of times over a given time horizon. Each knapsack has a constraint, being the overall budget for a specific time period. In its simplest form, the multiple knapsack problem then has the following form

$$x_{it} \in \{0, 1\}, \forall i \in 1, \dots N, t \in 1, \dots T, \tag{2}$$

$$\max \sum_{i=1}^{N} \sum_{t=1}^{T} v_{it} x_{it}, \tag{3}$$

$$\sum_{i}^{N} c_{it} x_{it} \leq b_t \ \forall t \in 1, \dots T, \tag{4}$$

$$\sum_{t}^{T} x_{it} \leq 1 \ \forall i \in 1, \dots N. \tag{5}$$

The decision variable x_{it} determines when the project will start. The first equation, the objective function specifies the total value v_{it} of projects i starting at time t. The first constraint specifies the total cost of projects i starting at time t, c_{it}. Notice that it is implicitly assumed that a project's cost is only incurred for one time period. In this chapter we consider items or projects whose costs extend over multiple time periods.

One must specify $c_{ti\tau}$, the cost of project i at time period t given that it starts at time τ. Thus decisions made for one time period or knapsack effect those of subsequent knapsacks. The last constraint specifies the item, in this case being the project start time, can only be selected once.

In the formulation of the optimisation problem for the *NITRO* application, additional constraints were required that reflected additional data on the types of projects, not just their cost profile and value. For example, project groups required the specification of temporal dependency. One project could not be included in the portfolio until another had commenced earlier.

The portfolio selection process also considered mutually exclusive project variants. For example, a project for procurement of one hundred office desks also has a variant where the number of desks to be purchased was fifty. Yet another variant could include the purchase of forty desks with extra storage draws. In the portfolio selection process only one such project variant would be considered for project selection.

Extra funding for new projects in part came from the cancellation or delay of extant projects. The portfolio selection process also placed constraints on the size of budget under- or over-spends.

Suppose $i \in \{1, \dots, |P|\}$ is a project drawn from the set P of projects. Time is represented as $t \in \{1, \dots, |T|\}$. Let $x_{it} \in \{0, 1\}$ be one if project i starts at time t and zero otherwise. Let $V : P * T \rightarrow \mathbb{R}$ be the value function. If project i starts at time t then its value is V_{it}. The cost at time t of project i starting at time τ is defined as $c_{ti\tau}$. Projects are classified as either extant, labelled E, or new, labelled N, with $E \cap N = \emptyset$ and $E \cup N = P$. Prior to portfolio optimisation extant projects have pre-defined timings specified by $\tilde{x}_{i\tau}, \forall i \in E$. The extant projects induce a fixed baseline budget which at time t is B_t where

$$B_t = \sum_{\tau=1}^{T} \sum_{i \in E} c_{ti\tau} \tilde{x}_{i\tau}. \tag{6}$$

This sum is known in advance as extant projects in Eq. 6 have known values of $\tilde{x}_{i\tau}$. The fixed overall budget is $\mathrm{B}_t = B_t + a_t$ where a_t are yearly additions or subtractions to the fixed baseline budget. Once the prior timing of extant projects E is used to calculated the fixed overall budget, their timing decisions, $\tilde{x}_{i\tau}$ are discarded. Projects in E are then placed in the overall project pool P to have their inclusion and timing recalculated through optimisation.

Before defining the overall set of equations used to calculate the portfolio, recall that budgets are subject to negotiation for minor changes. The objective function not only attempts to define a portfolio that maximises value but also control the level of budgetary under- and over-spend. To do this the decision variables $\epsilon_t^{\text{over}}, \epsilon_t^{\text{under}} > 0, \forall t \in \{1, \dots, T\}$ are defined to represent the total over- and under-spend from the budget, also termed budgetary slack. The linear constraints imposed are

$$\sum_{i \in P} c_{ti\tau} x_{i\tau} \leq \mathrm{B}_t + \epsilon_t^{\mathrm{over}} \text{ and} \tag{7}$$

$$\sum_{i \in P} c_{ti\tau} x_{i\tau} \geq \mathrm{B}_t - \epsilon_t^{\mathrm{under}}, \forall t \in \{1, 2, \ldots, T\}. \tag{8}$$

There is an immediate interplay between x_{it} and $\epsilon_t^{\{\mathrm{under,over}\}}$. If the slack is large there is more scope for both the number of projects in the portfolio and their movement in time. The bound on this interplay is set in the formation of the objective function which will be discussed below. It should be noted for optimal solutions, $x_{it}^{\star}, \epsilon_t^{\star,\{\mathrm{under,\ over}\}}$ that

$$\epsilon_t^{\star,\mathrm{over}} = \max\Big(\sum_{\tau=1}^{T} \sum_{i \in P} c_{ti\tau} x_{i\tau}^{\star} - \mathrm{B}_t, 0\Big), \tag{9}$$

$$\epsilon_t^{\star,\mathrm{under}} = \max\Big(\mathrm{B}_t - \sum_{\tau=1}^{T} \sum_{i \in P} c_{ti\tau} x_{i\tau}^{\star}, 0\Big). \tag{10}$$

The balance between maximising overall portfolio value versus budget accuracy is governed by taking a convex combination of the value term $\sum_{i \in P} \sum_t V_{it} x_{it}$ and the negative of the total budget under and over spend $\sum_t (\epsilon_t^{\mathrm{under}} + \epsilon_t^{\mathrm{over}})$.

Let $\xi_l = \{i_1, i_2, \ldots, i_{\xi_l} \in P\}$, $l \in \{1, \ldots, L\}$ denote a group of projects that are mutually exclusive in the portfolio, which implies

$$\sum_{i \in \xi_l} \sum_{t=1}^{T} x_{it} \leq 1, \forall l \in L. \tag{11}$$

Thus in the portfolio at most one project in ξ_l can be chosen. If the right hand side of (11) is equality, one and only one project in ξ_l *must* be in the portfolio.

Let us define pairs of projects that have some temporal inter-dependency. One may constrain pairs of projects such that one must start before another, if at all. Consider two projects $i, j \in P$ and suppose that project i is constrained to start before project j by some time period s if either projects start at all. The constraint for this is simply

$$x_{i,t-s} \geq x_{jt} \ \forall t \in \{s+1, s+2, \ldots, T\}. \tag{12}$$

Let $P_k(i_{k1}, i_{k2}, s)\, k \in \{1, \ldots, K\}$ be a pairs of projects constrained such that if project i_{k2} starts at time t, project i_{k1} must start at time $t - s$.

Clearly, prior to any optimisation, a logical check of the constraints for consistency must occur. The imposed constraints have the potential to be contradictory. For example, two projects cannot be both mutually exclusive whilst one starts before another. These are contradictory constraints. Using the notation described above, suppose projects $i_1$and $i_2 \in \xi_l$, that is either project i_1 or i_2 is chosen but not both.

Clearly the additional constraint that i_1 starts before i_2, that is $i_1, i_2 \in P_k(i_1, i_2)$ would force the optimisation result to be infeasible.

Given the assignment of value is so subjective, force designers should have the option of mandating the initiation of a project over the decision horizon or at a specific point of time. If $M \subsetneq P$ is the set of mandated projects which are initiated over the planning horizon, and $M(\tau)$ is the set of projects mandated to start at τ, then the corresponding constraints are simply

$$\sum_{t=1}^{T} x_{it} = 1, \forall i \in M, \tag{13}$$

$$x_{i\tau} = 1, \forall i \in M(\tau). \tag{14}$$

Having discussed the constraints, this leads us to the entire set of equations that are optimised to generate a Defence portfolio,

$$\max_{x_{it}, \epsilon_t^{\text{under}}, \epsilon_t^{\text{over}}} \alpha \left[\sum_{i \in P} \sum_{t=1}^{T} V_{it} x_{it} \right] - (1 - \alpha) \left[\sum_{t=1}^{T} (\epsilon_t^{\text{under}} + \epsilon_t^{\text{over}}) \right] \tag{15}$$

subject to

$$x_{it} \in \{0, 1\}, \quad \forall i \in P, \quad \forall t \in \{1, \ldots, T\}, \tag{16}$$

$$\epsilon_t^{\text{under}}, \epsilon_t^{\text{over}} \geq 0 \;\; \forall t \in \{1, \ldots, T\}, \tag{17}$$

$$\alpha \in [0, 1], \tag{18}$$

$$\sum_{t=1}^{T} x_{it} \leq 1, \quad \forall i \in P, \tag{19}$$

$$\sum_{t=1}^{T} x_{it} = 1, \quad \forall i \in M, \tag{20}$$

$$\sum_{i \in \xi_l} \sum_{t=1}^{T} x_{it} \leq 1, \forall i \in \xi_i, \quad \forall l \in \{1, \ldots, L\}, \tag{21}$$

$$\sum_{i \in P} \sum_{\tau=0}^{T} c_{ti\tau} x_{i\tau} \geq \mathrm{B}_t - \epsilon_t^{\text{under}}, \quad \forall t \in \{1, \ldots, T\}, \tag{22}$$

$$\sum_{i \in P} \sum_{\tau=0}^{T} c_{ti\tau} x_{i\tau} \leq \mathrm{B}_t + \epsilon_t^{\text{over}}, \forall t \in \{1, \ldots, T\}, \tag{23}$$

$$x_{i_k, t-s} \geq x_{j_k t} \;\; \forall t \in \{s+1, s+2, \ldots, T\}, i_k, j_k \in P_k(i_k, j_k), k \in \{1, \ldots, K\}. \tag{24}$$

The parameter α controls the trade-off between maximising the portfolio score and maintaining budget. As $\alpha \rightarrow 1$, the portfolio value is optimised, however there may be significant deviations away from the budget as is measured by $\sum_t^T (\epsilon_t^{\text{under}} + \epsilon_t^{\text{over}})$. As $\alpha \rightarrow 0$, it is the budget integrity that is key issue.

3.1 Inherent Solution Challenges

There are $|P|T + 2T$ decision variables in the linear program. There are also $|P| + |M| + |L| + 4T$ constraints without temporal inter-dependency. For one hundred projects, over a ten year time horizon, this amounts to 1020 decision variables and at least 140 constraints. Clearly the linear program quickly grows in complexity as the number of projects increases.

The full portfolio optimisation problem is a mixed-integer program as x_{it} are binary and ϵ_t^{over} and $\epsilon_t^{\text{under}}$ are real-valued. As such this problem is NP-hard. To solve the problem, the solution space is first relaxed, so $x_{it} \in [0, 1]$. Then either the SME chooses integer values from the relaxed solution or the technique of branch and bound [30] can be applied to find the optimal binary-valued solution. This approach, first integer programming relaxation, then the application of branch and bound, was originally proposed by pioneering female operations research scientists, Ailsa Land from the United States and the Australian Alison Doig (latter Alison Harcourt [3]), and refined by Dakin in 1965 [7]. Unsurprisingly the algorithm is termed the Land-Doig-Dakin algorithm.

In practice for large problem spaces, the application of the branch and bound algorithm post-relaxation was too slow to meet the requirements of the SME. Rather such experts choose to decide themselves the starting times of projects where the relaxed solution was not binary. These considerations will be discussed further in Sect. 5.

Readers should note there are a number of simple transformations to the objective function (15) that may be implemented, given additional information. For example, it was assumed project value was assigned at the commencement of the a project. Instead, one could assign a specific time a project makes the transition from an acquisition and capability development stage to a fully operational, sustainment phase. Let δ_i be the time taken for project i, from its instigation, to move from acquisition to sustainment. Furthermore, let the utility of project i once it has reached sustainment phase at time t be U_{it}, and the binary decision for whether to include project i when it reaches sustainment phase be y_{it}. Then the portfolio utility term of Eq. (15) will be

$$\sum_{i \in P} \sum_{t=1}^{T} U_{it} y_{it} = \sum_{i \in P} \sum_{t=1}^{T} U_{it} x_{i,t-\delta_i}. \tag{25}$$

This simple transformation implies that the same binary program holds when utility is assigned at the time of capability.

Another issue is the finite time horizon. Projects scheduled at times closer to the budget horizon T are subject to fewer yearly budgetary constraints. With a strict budget constraint, the optimised portfolio had projects scheduled latter in the planning time horizon, or "shifted to the right." With the introduction of budget under- and over-spends the shift of projects to the right was decreased.

To counter the inherent bias of a finite time horizon, we introduced cumulative valuing of projects across the time horizon. Recall, each project $i \in P$ is assigned a value at time $t \leq T$ termed V_{it}. The value here is interpreted as a contribution to the portfolio from project i *starting at* time t. However, value may have other interpretations. For example, value may be interpreted as *cumulative benefit.* The cumulative benefit may be the value at project instigation plus value imparted from project continuity. Thus the cumulative benefit may be defined as

$$\hat{V}_{it} = \sum_{\tau=t}^{T} V_{i\tau}. \tag{26}$$

By definition $\hat{V}_{it} \geq \hat{V}_{i,t-1}$ thus there will be a tendency in the optimised portfolio to move projects earlier in time.

An alternative approach could be to introduce a time discount for the value of projects in the future [18]. The time discounting of rewards is a significant idea of research psychology and behavioural economics [13, 45]. Some researchers have proposed exponential discounting of rewards or value where a modified reward function may take the form $\tilde{V}_{it} = e^{-\beta t} V_{it}$. Here β governs the discounting rate. An alternative is the hyperbolic discounting of value $\tilde{V}_{it} = \frac{V_{it}}{1+kt}$ where k is the discount factor [29].

To allow for extra flexibility and ensure feasibility, a total budget under-spend or over-spend can be allowed. Additional constraints can be added so that some subgroup of projects must start before some specified time period, if they start at all. This can be achieved by adding extra constraints into the mixed-integer program. Consider a subset of projects $P_\tau \subset P$ that must start prior to some specified time τ. One adds the constraint

$$\sum_{t=\tau+1}^{T} x_{it} = 0, \quad \forall i \in P_\tau. \tag{27}$$

Having described the mathematical structure of the optimisation problem, its implementation as a software application is explained in the following section.

4 Implementation in Microsoft Excel®

The implementation of the algorithm was done in Microsoft Excel® using the What's*Best!*® linear programming solver. In this section we discuss this implementation.

Figure 3 illustrates the flow of information in the NITRO application. The costing and decision data are combined to calculate the total budget (determined from $\tilde{x}_{i\tau}$), then the portfolio determined yearly cost (determined from $x_{i\tau}$). The yearly cost is then applied to calculate the values of $\epsilon_t^{\{\text{under,over}\}}$. The decisions, budget, and yearly cost are fed into Excel cells that specify the linear programming constraints. The objective function is also determined in an Excel cell then the What's*Best!* application is run to determine the optimal portfolio decision.

The costing data is represented as a cost block of cells in Excel as is the binary decision data. The cost cells of data are known as the cost block, **C**, which is a $|P| \times T$ range of cells specifying the values of c_{ti1}. Thus the columns of the cost block specify the expenditure per year, assuming each project starts at year one. Similarly, the decision block **X** is a $|P| \times T$ range of cells that specifies the values of $x_{i\tau}$. Here, the columns of the decision block specify which calendar year the project in a commensurate row starts in. We also define yearly cost blocks $\mathbf{C}_\tau$ for each possible project starting year $\tau \in \{1, \ldots, T\}$, over $|P| \times T$ cells. The τth yearly cost block for project i at calendar year t is defined as

$$\mathbf{C}_\tau(i, t) = c_{ti\tau} x_{i\tau}, \forall i \in P, \forall t \in \{1, \ldots T\}. \tag{28}$$

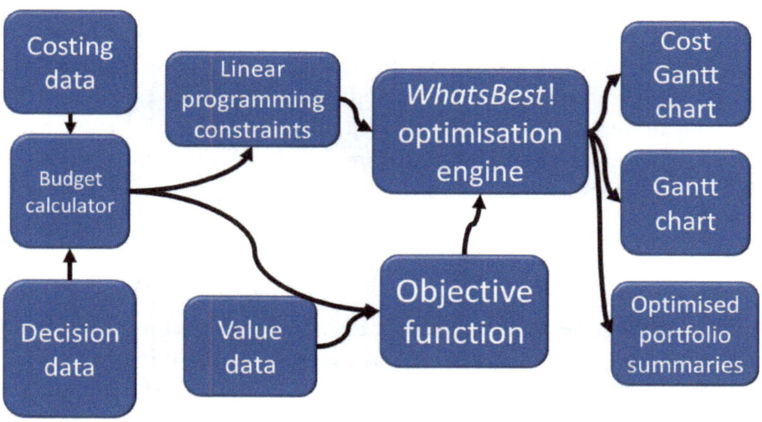

Fig. 3 This figure illustrates the flow of data, project costs $c_{ti\tau}$ and decisions x_{it} through to the budget. The budget combined with the optimisation constraints and objective function serve as inputs to the What's*Best!*® optimisation engine

The columns of the τth yearly cost block specify the annual costs of projects starting at time τ. Intuitively the columns of $\mathbf{C}_\tau$ contain the values of $c_{ti\tau}$ only if project $i \in P$ starts at time τ. These values are identical to c_{ti1} however they are shifted to the right by τ cells. If a project $j \in P$ does not start at time τ then the jth row of the yearly cost block $\mathbf{C}_\tau$ has all values equal to zero. Summing the costs across the distinct yearly cost blocks enables one to calculate the overall yearly cost and compare such values to the yearly budget. The yearly cost blocks take advantage of the fact that NITRO uses absolute, not inflation-adjusted, prices. This implies that $c_{t+\tau-1,i,\tau} = c_{ti1}$.

Typical linear programming solvers require users to format the linear program into the standard form before running the appropriate algorithm. As What's*Best!* is based on Excel, it requires the use of the **R1C1** formula notation to calculate the linear programming constraints. For simple problems the use of Excel formulas to specify the structure of the linear program is advantageous, as one can simply apply Excel functionality in the calculation of the constraints. For simple problems, formulating the What's*Best!* linear programming constraints is highly visual. However the calculation of the portfolio yearly cost, as a function of the decisions $x_{i\tau}$ required explicitly enumerating every project to start in every year of the time horizon and coding an Excel function to calculate the yearly cost blocks.

The calculation of the yearly cost blocks relied on both the size of the cost and decision blocks and also the overall geometry of the Excel spreadsheet. This geometry is illustrated in Fig. 4. The boundary of this figure represents the border of the Excel spreadsheet. In blue are the decision block **X**, the cost block **C**, and the yearly cost blocks $\mathbf{C}_1, \mathbf{C}_2, \ldots$. The final yearly cost block, $\mathbf{C}_T$ is not illustrated in the figure. The number of columns between the Excel spreadsheet border and the first entry in the decision block **X** is G_L. The number of columns between the left Excel border (column A) and the cost block is $G_L + |P| + G_{XC}$, whilst the number of rows between the top Excel border (row 1) and the decision and cost blocks is G_T. These

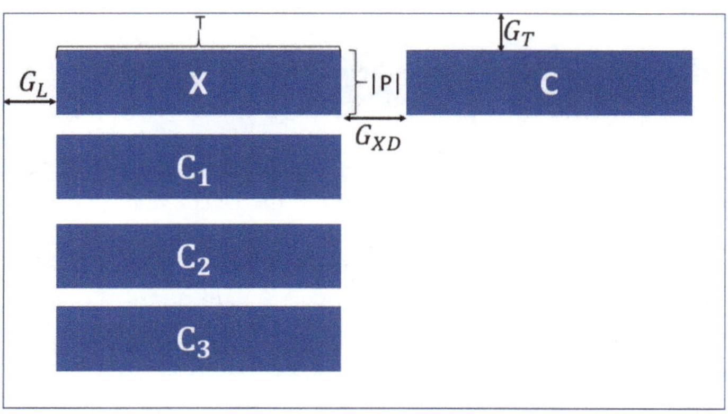

Fig. 4 This figure illustrates the geometry of the Excel® spreadsheet to calculate, project costs and decisions through to the budget

parameters determine the iterative **R1C1** formula for the calculation of the yearly cost blocks.

For the τth yearly decision block, the first potentially non-zero cell is $C_\tau(1, \tau)$ as all projects in this block start at τ. Assuming a one row gap between each yearly cost block, the Excel cell position of $C_\tau(1, \tau)$ in Visual Basic notation is

$$\text{Cells}(G_T + \tau(|P| + 1) + 1, G_L + \tau). \tag{29}$$

In Visual Basic this cell is selected and a formula is applied that contains relative references to the decision and cost blocks, **X** and **C.** Let

$$r = -\tau * (|P| + 1) \tag{30}$$

$$w = G_W + \tau \tag{31}$$

$$x = T + G_{XC} - (\tau - 1), \tag{32}$$

then the **R1C1** formula placed in the cell specified by Eq. (29) is

$$\text{ActiveCell.FormulaR1C1 ="R[" \&}r\text{\&"]C"\&}w\text{\&"*R["\&}r\text{\&"]C["\&}x\text{\&"]"}. \tag{33}$$

Notice the formula in Eq. (33) contains an absolute reference to the projects that start at time τ, which is the term C"&w&.

To complete the population of costs for projects that start in the τth year, the autofill functionality of Excel is applied. If one defines

$$a_1 = G_T + \tau(|P| + 1) \tag{34}$$

$$a_2 = G_L + \tau \tag{35}$$

$$a_3 = G_L + T \tag{36}$$

then the autofill command for the first row of the τth yearly cost block is

$$\text{Selection.AutoFill Destination:} = \text{Range(Cells(}a_1, a_2\text{), Cells(}a_1, a_3\text{))},$$
$$\text{Type:} = \text{xlFillDefault}.$$

A similar Excel autofill formula completes the subsequent rows of the τth yearly cost block. Figure 5 illustrates the application of the Excel yearly cost block macro. In this example, there are three calendar years and three projects. Project one starts in year one, project two in year three and project three in year two. Figure 5 shows the value of the yearly cost blocks and the how these are summed to find the overall yearly portfolio cost.

Having explained the calculation of the yearly portfolio cost the linear constraints and objective function are formed. What's*Best!* has inequality-equality operators $\leq$, $\geq$, and $=$ that are applied into an Excel cell. The left hand side cell specifies

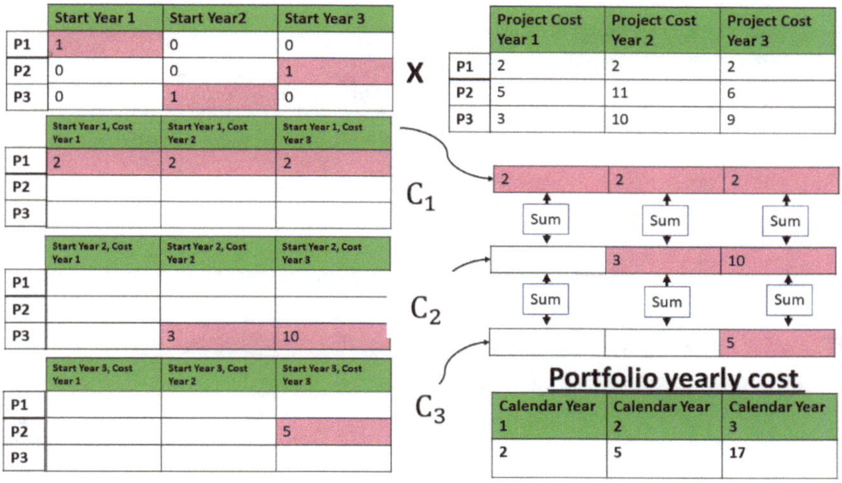

Fig. 5 Example of the application of Excel **R1C1** formula in the calculation of the overall portfolio yearly cost. The top left table is the decision block which specifies the times projects start in the optimised portfolio. The top right table specifies the cost block project costs per year, assuming the projects start at year one. These blocks are combined to form the yearly cost blocks which specify the costs of a project assuming they start at a specified year. In turn, these are summed to calculate the overall portfolio cost

the value, then the inequality-equality constraint followed by the right hand side constraint value.

The value function is calculated by taking the element sum of the Hadamard product of the value block which is the $|P| \times T$ range of cells that specify V_{it} which the decision block **X**.[1] This is done using the Excel SUMPRODUCT function. This is combined through convex combination with the summed values of $\epsilon_t^{\text{under}}$ and ϵ_t^{over} to find the overall portfolio objective function.

In terms of visualisation, the Excel's conditional rule based colouring of cells was extensively applied to display the results of the optimisation.

For example, a simple cell based rule applied to the decision block coloured cells to formed a Gantt chart visualisation of the optimised project schedule. The Gantt chart is a $|P| \times T$ range of cells. For project $i \in P$ and $t \in \{1, \ldots, T\}$ the Gantt chart value of the cell is $\sum_{\tau=1}^{t} x_{i\tau}$. Thus for each project, the cells have a yearly value of zero prior to the commencement of a project and one thereafter. The cells in the Gantt range that have value one are given a colour green, otherwise a default colour of white is set. The rule implemented in **R1C1** notation checks with an OR rule the value of the decision from **X** or the value of cell of the same row, in the antecedent column. This rule is

[1] The Hadamard project between two matrices of the same size **A** and **B** denoted $\mathbf{A} \odot \mathbf{B}$ is the term by term product of the two matrices $(\mathbf{A})_{ij}\ (\mathbf{B})_{ij}$.

Decision Block **X**

	Start Year 1	Start Year 2	Start Year 3	Start Year 4
Project 1	0	1	0	0
Project 2	0	0	1	0
Project 3	1	0	0	0
Project 4	0	0	0	1

Cost Block **C**

	Project Cost Year 1	Project Cost Year 2	Project Cost Year 3	Project Cost Year 4
Project 1	2	3	3	5
Project 2	2	7	12	14
Project 3	4	4	0	0
Project 4	6	3	2	2

Project Gantt Chart

	Start Year 1	Start Year 2	Start Year 3	Start Year 4
Project 1	0	1	0	0
Project 2	0	0	1	0
Project 3	1	0	0	0
Project 4	0	0	0	1

Cost Gantt Chart

	Start Year 1	Start Year 2	Start Year 3	Start Year 4
Project 1	0	2	3	3
Project 2	0	0	2	7
Project 3	4	4	0	0
Project 4	0	0	0	6

Fig. 6 Illustration of the simple visualisation of projects and project costs. The Gantt chart and dollar Gantt charts took as input the optimised portfolio, through the decision block **X** and the cost block **C**. These Gantt charts were displayed in different worksheets of the NITRO application

if(OR("Sheet"!R[r]C[c]=1, RC[-1] =1),Cells.Color=Green, Cells.Color=White) (37)

where r and c are the relative row and column references to the Excel worksheet labelled "Sheet" which contains the decision block **X**.

A dollar Gantt chart was formed by using the Excel SUMIF function. The yearly cost blocks contain the cost values of each project offset appropriately. Each project $i \in P$ in cost block $\mathbf{C}_\tau$ has

$$\mathbf{C}_\tau(i,t) = \begin{cases} c_{t,i,\tau} x_{i\tau} = 0, & t < \tau, \\ c_{t,i,\tau} x_{i\tau} = c_{t-\tau+1,i,1} x_{i\tau}, & t \geq \tau. \end{cases} \tag{38}$$

Let r_f be defined as the first row of all the yearly cost blocks and let r_e be the endmost row. The yearly cost blocks are assumed to be placed in an Excel labelled worksheet. Each cost block is labelled with rows as projects $i \in \{1 \ldots P\}$ in column 1. The value of project i in the dollar Gantt chart at time t is thus

$$\text{SUMIF}(\text{R}r_f\text{C}[1]\text{:R}r_e\text{C}[1],\, i,\, \text{R}r_f\text{C}[t]\text{:R}r_e\text{C}[t]). \tag{39}$$

That is, the cost in each yearly cost block is summed only if the corresponding row in column 1 is labelled with project i. Again the conditional rule for colouring was applied. Cells that had non-zero costs were coloured red, with smaller values having greater colour transparency. The Gantt chart and the dollar Gantt chart are illustrated in Fig. 6.

5 Performance and Budget-Value Trade-Offs

The computational complexity of the NITRO application was briefly discussed in Sect. 3.1. In this section we present some statistics on the performance of the NITRO application and present some results on the trade-off between the maintaining a strict budget ceiling and the overall value of the portfolio.

A primary *raison d'etre* for the development of the NITRO application was to enable SMEs to develop and refine optimised portfolios in real time. SMEs considered a process of data entry followed by off-line calculation unsuitable for use in workshops where iterated generation of portfolios would occur as part of the decision making process. Furthermore, the NITRO application ran on limited computing resources. For example, the performance statistics presented in this paper were based on a Dell® laptop with an Intel® Core™ i7, 8th generation processor. Typically the number of projects that were considered in the portfolio ran into the hundreds.

Given these constraints it is worth commenting on the practicality of using mixed-integer programming to conduct the portfolio optimisation. We used data generated using the procedures in [20] from the University of New South Wales (UNSW) School of Engineering and Information Technology [19]. This data contains simulated cost profiles for five hundred projects. Project durations and total cost were independently sampled from a multivariate lognormal distribution found to model Defence project expenditures [60]. The parameters for the distribution were derived from publicly available investment plans [51]. Annual expenditures were based on Weibull distributions as described in [61] over forty calendar years. Each project is given a random value using the generation procedure in [20].

The branch and bound algorithm was applied to the portfolio optimisation with the UNSW data set. Sensitivity analysis over the time to optimality was conducted for $\{5, 10, 15, 20, 25, 30, 35, 40\}$ projects, where each project has 16 possible start years, which means the number of decision variables in the integer linear programming problem was $16|P|$. No slack variables $\epsilon_t^{\{\text{over,under}\}}$ were implemented. The yearly budget $\mathbf{B}_t$ was assumed to be $12|P| + (t-1)|P|$, for $t = \{1, 2, \ldots 40\}$. Such a budget ceiling was found by experimentation to admit a reasonable number of projects in the portfolio, as a function of the total number of candidate projects $|P|$. For comparative purposes, time to optimality was compared using the Excel based What's*Best!* solver and the Python®-based Gurobi® optimisation package. The results of the analysis are shown in Fig. 7.

Figure 7 demonstrates the performance of the Gurobi package is superior to that of What's*Best!* in terms of time to optimality. However, given the performance criterion of real-time optimisation, the binary-programming performance of both Gurobi and What's*Best!* are insufficient. With only forty projects the Gurobi package took 22524 s (6.2 h) whilst the What's*Best!* package took 109209 s (1.2 days) to complete. Large organisations typically have significant numbers of candidate projects to be considered for portfolio inclusion, therefore another approach to portfolio optimisation was required.

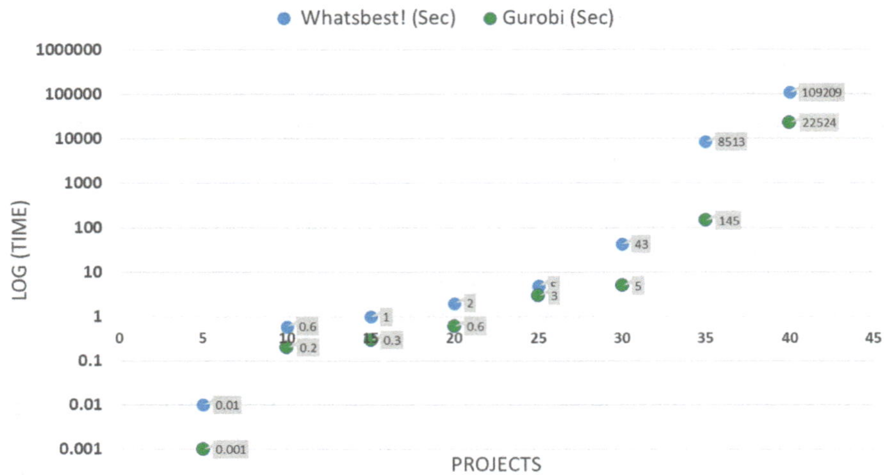

Fig. 7 Here we compare the performance of What'sBest! implemented in Excel with Gurobi solver implemented in Python using the same processor. The sensitivity analysis was conducted from 5 to 40 projects. The time to optimality is shown on a logarithmic scale

As discussed earlier the binary program was relaxed. Thus solutions were considered over the real valued interval [0, 1] not {0, 1}. The time to optimise the relaxed solution, using the What's*Best!*, PuLP-CBC (or just CBC), and Gurobi linear program solvers is shown in Fig. 8. The What's*Best!* solver records the optimisation time in whole seconds whereas the Gurobi and CBC solvers record time in fractions of seconds.

One can observe that for small numbers of projects, the What's*Best!* optimisation was generally slower than that of the CBC and the Gurobi solvers. A possible reason for this comes from the observation that What's*Best!* is an Excel add-in and thus must extensively pre-process the data and constraints from that of the Excel format to that required for the background solving engine, which is the LINDO® application programming interface (API) [31]. It is the authors' understanding that the LINDO API can also be used in conjunction with other programming languages.

For larger numbers of projects What's*Best!* performed just as well as Gurobi and CBC. It could concluded that What's*Best!* is a comparable solver to that of other commercially available solvers. In a comparative review of linear programming solvers, Kronqvist showed that the LINDO solver could only optimise a smaller number of problem instances with specific non-linearities when compared to other commercially available solvers [28].

Nonetheless, What's*Best!* has an advantage in that it is implemented in an environment familiar to many in the Australian DoD, Excel.

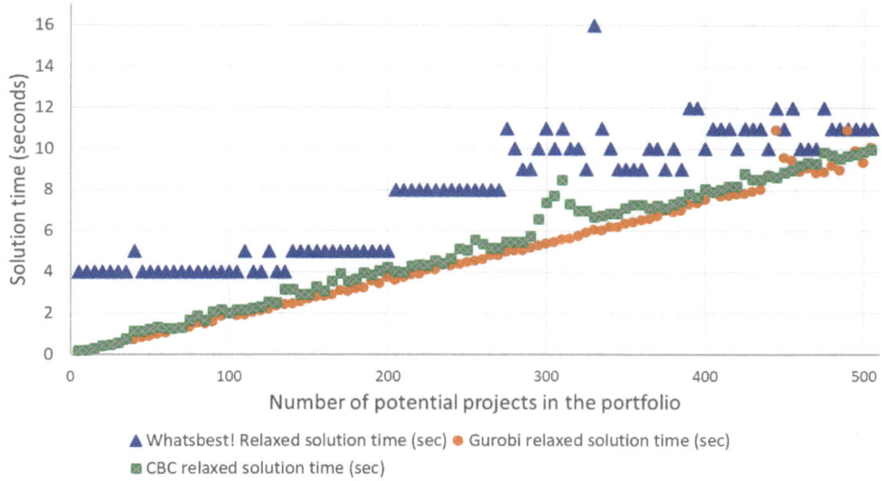

Fig. 8 Here we compare the optimisation processing time of What's*Best!* implemented in Excel with the Gurobi and CBC solver implemented in Python using the same processor for the relaxed problem. The number of projects considered in the portfolio, from 5 to 505 in increments of 5 is displayed on the horizontal axes. The time to optimality is shown on the vertical scale in seconds

5.1 Relaxation

Section 3.1 alluded to the possibility that SMEs could choose between non-binary variables in the relaxed portfolio optimisation program. In considering this approach there are a number of obstacles. First, the number of non-binary solutions may be large. Second, depending on the value of the non-integer solutions, it may not be clear as to the best heuristic to apply. For example, consider three heuristics applied to the optimised relaxed solution x_{it}^*, where $\sum_t x_{it}^* > 0$ for project i. One, if $\sum_t x_{it}^* < 1$, one could eliminate the project from the portfolio altogether, setting the heuristic solution, x_{it}^H for project i as $\sum_t x_{it}^H = 0$. Another approach is to find $t_H = \arg\max_t\{x_{it}^*\}$ and set $x_{it_H}^H = 1$, that is setting the maximum non-zero solution for project i to be one. The issue here is the possibility of multiple ties in the value of x_{it}^*. The final heuristic considered is rounding. This is one setting $x_{it}^H = [x_{it}^*]$, where $[x]$ is zero if $x < 0.5$, and one otherwise.

In the analysis that follows this last heuristic was implemented, because it is both intuitive and minimises the possibility of tied decisions for a project.

The first issue, that of a large number of non-binary solutions would make the process of a SME choosing binary replacements the most pressing, as decisions on the inclusion and start times of many projects may become untenable. To test this viability the structure of the relaxed solutions was analysed using the UNSW project costs-value data set. A sensitivity analysis over different numbers of potential projects for inclusion in the portfolio was conducted. The number of binary versus non-binary solutions was recorded along with the objective function of the relaxed

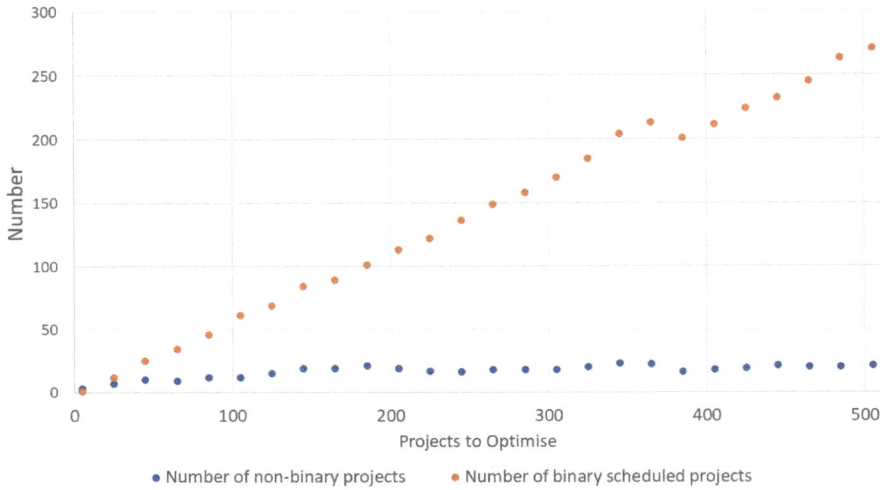

Fig. 9 Sensitivity analysis of binary versus non-binary projects generated by the relaxed portfolio when using the Gurobi solver

solution $\sum_i V_i x^*_{it}$ (in the data set the project values are time independent) and the objective function of the rounding heuristic $\sum_i V_i x^H_{it}$. Figure 9 plots the results for 5 to 505 projects in increments of 20 projects. The optimisation engine used was the Gurobi solver.

Figure 9 shows that as the number of candidate projects in the portfolio increases, the projects with non-binary solutions increases at a much slower rate than those of the binary solutions from the optimised portfolio using the Gurobi solver. For example, with 505 projects there are 21 projects with non-binary solutions compared to 271 binary projects. Thus for a large number of projects, the potential to have the SME either choose an appropriate project timing or implement a heuristic such as rounding is feasible. Having said that, it is important to emphasise here that the ratio between binary and non-binary projects can vary depending on the algorithm used. For example, PuLP CBC relaxed solution has more non-binary projects when compared to Gurobi, and because of this the portfolio value after applying the rounding heuristics method is lower than Gurobi solution. Although the solution time for these two solvers is very similar, as show in Fig. 8, the Gurobi solver has a higher portfolio value solution after rounding the decision variables. This becomes clear from Fig. 10 showing that the CBC asymptotic value is a bit higher than that of Gurobi and What's*Best!*.

Furthermore, it can be shown that the relaxed solution converges to the rounding heuristic solution as the number of candidate projects grows. If $O_R^{|P|}$ is the objective function of the relaxed solution and $O_H^{|P|}$ that of the rounding heuristic, one observes empirically that

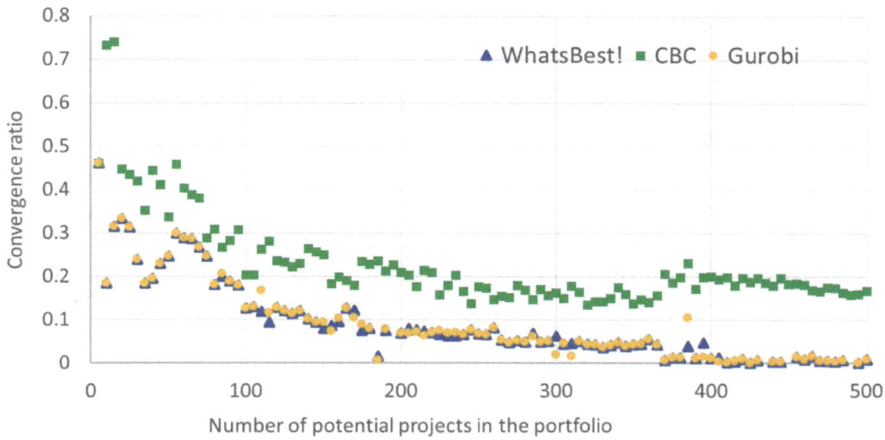

Fig. 10 Ratio of the relaxed solution minus the rounding heuristic solution divided by the relaxed solution, that is $1 - \frac{O_H^{|P|}}{O_R^{|P|}}$. The project numbers considered for inclusion into the portfolio range from 5 to 505 in increments of 20 projects. A comparison has been made between the What's*Best!*, CBC, and Gurobi solvers

$$1 - \frac{O_H^{|P|}}{O_R^{|P|}} \to 0 \text{ as } |P| \to \infty. \tag{40}$$

This observation can be seen from the following Fig. 10.

The CBC solver uses the cut-and-branch algorithm to solve the relaxed problem [12], which uses cutting planes to find the optimal solution [11]. The use of cutting planes may induce a greater "spread" of solutions where $x_i^R < 0.5$. Once rounded the heuristic optimal will be less than the optimal using the branch and bound algorithm alone, implying the convergence ratio may be higher.

It should be noted that studies of the difference between the relaxed solution and the optimal integer linear program have been done in the past. For example for the standard knapsack problem, Goldberg and Marchetti-Spacamella showed that for N decision variables, the difference between the relaxed solution and the optimal solution is $\mathcal{O}\left(\frac{\log^2(N)}{N}\right)$ [16].

Puchinger et al. [43] highlighted the result from Kellerer that there exists a relaxed solution to an integer knapsack problem with at most $\min\{n, m\}$ fractional values, where n is the number of decision variables and m the number of constraints [26].

In [43] a computational analysis was conducted on the differences between the optimal binary programming solution, x_j^* and the optimal relaxed solution x_j^R, $J = \{j = 1, \ldots, n\}$, for knapsack problems, of moderate complexity, where the optimal solution could be computed in reasonable time. Puchinger et al. considered the Hamming distance

$$\Delta LP = \sum_{j \in J} |x_j^* - x_j^R|. \tag{41}$$

The authors considered three sets of decision variables, the subsets $J_{\text{int}} \subseteq J$ where x_j^R are integral, $J_{\text{frac}} \subseteq J$, where x_j^R are fractional and the Hamming distance where the relaxed solution is rounded, $[x_j^R]$. Problems where the number of decision variables $n = 100$ or 250 with constraint numbers $m = 5$ or 10 were considered.

The Hamming distance between the relaxed and optimal solution was found to have decreased as the number of decision variables increased. Indeed, for the rounding decision heuristic, only 6% of 100 decision variables deviated away from the optimal. Approximately 3% of 250 decision variables deviated away from the optimal binary program. Thus one sees a similar but not identical empirical observation in the literature regarding the convergence of the relaxed solution to the optimal binary solution. At the time of writing, a detailed proof of these empirical observations is not available in the literature.

5.2 *Value-Slack Trade-Offs and the Issue of Sensitivity*

Recall the original equations that constituted those of the *NITRO* application included the introduction of budget slack, which allows flexibility in budget under or over spends. The introduction of budget slack allows SMEs to trade-off budget accuracy with value. Recall that as $\alpha \to 1$ the portfolio value takes primacy over budget slack. As $\alpha \to 0$ minimising budget slack takes primacy. In this section, we show that there is indeed a trade-off between value and budget accuracy through empirical analysis. Figure 11 displays a sensitivity analysis as function of α. The maximum number of admissible projects in the portfolio is 20. Values and costs are taken from [20].

As is expected the number of projects included in the optimised portfolio increases as $\alpha \to 1$. Budget slack in this situation is not considered important and as all projects have positive value, the optimiser adds all twenty to the portfolio. It should also be observed that there is a discontinuous jump in overall portfolio value as $\alpha \to 0.5$.

The analysis of the continuity and structure of optimal solutions as parameters alter falls under the topic of parametric programming. Martin [34] was able to show that if a linear program is parameterised by some real valued parameter t, meaning that the objective function and constraint coefficients were dependent on t, then there was no assurance that the optimal value $V^*(t)$ was continuous as a function of t.

Martin has shown that the optimal solution $V^*(t)$ was upper semi-continuous as a function of t. Such a condition is weaker than a continuity, and is mathematically expressed as $\limsup_{t \to t_o} V^*(t) \leq V^*(t_o)$. Intuitively the optimal solution may "jump" as a function of parameter changes.

Hadigeheh and Terlaky [17] introduce the idea of *invariant support sets* or regions in the optimisation parameter space[2] where the optimal portfolio does not change.

[2] Which in the context of this paper are the project values and costs.

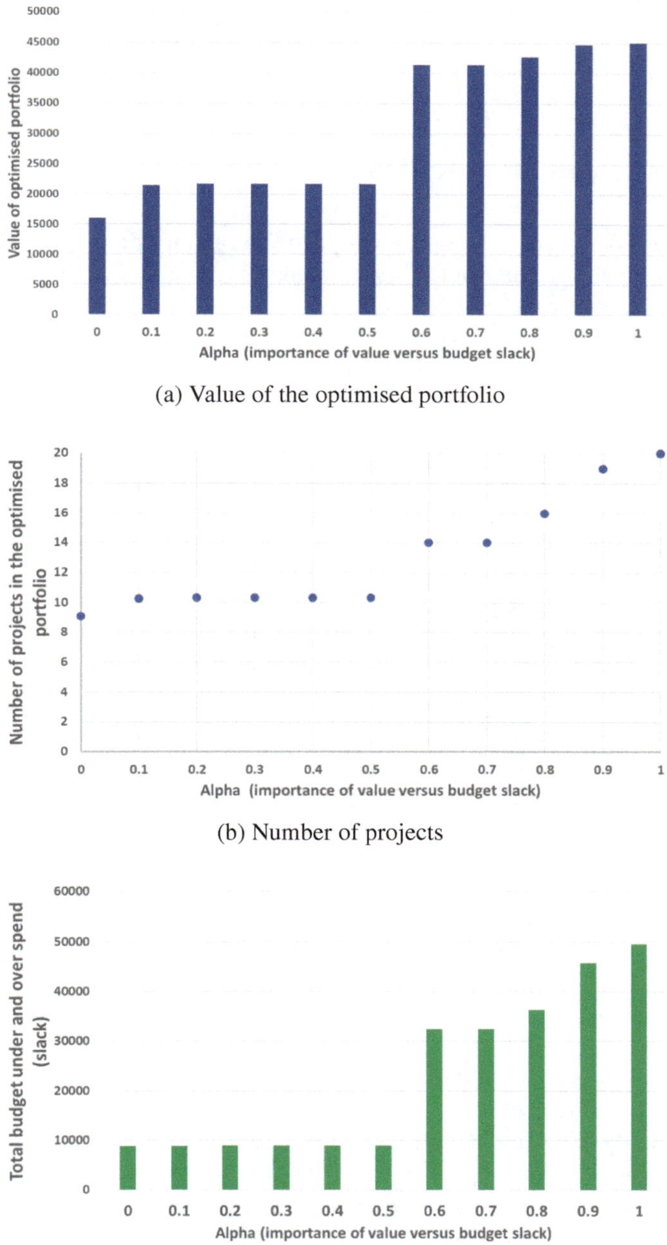

(a) Value of the optimised portfolio

(b) Number of projects

(c) Total budget slack

Fig. 11 The first graph (11a) shows the overall portfolio value as a function of α. The second graph (11b) displays the number of projects admitted to the optimised portfolio as a function of α. Finally, the third graph (11c) shows the total budget slack as a function of α

Understanding the "size", through some metric of these invariant support sets will be a crucial research endeavour.

6 Discussion and Future Work

This chapter commenced by discussing portfolio optimisation in a Defence context and described the DoD capability development process. Portfolio optimisation occurs after an extensive series of activities aimed at defining the pool of current and future projects worthy of consideration and understanding the context of project development, project values, costs, dependencies and timing constraints. Only after these precursor activities are completed can one consider the process of portfolio optimisation.

A mathematical formulation of the Defence portfolio optimisation was formulated in Sect. 3. Its essential structure is that of a multiple knapsack problem where the size of the knapsacks are the yearly budget constraints. The optimisation problem implementation in the application environment of Excel and What's*Best!* was described in Sect. 4. Section 5 covered a comparative computational performance of the What's*Best!* optimisation application in the Excel environment with Gurobi and CBC implemented in the Python environment. While both Gurobi and CBC were significantly faster than What's*Best!* in binary optimisation, for significant problem sizes the requirement for rapid optimisation in the order of seconds implied problem relaxation was inevitable.

Section 5.1 covered the issue of relaxation. Relaxation of the linear program allowed the application of such algorithms as interior point methods, but came at the cost of non-integer solutions. Here the What's*Best!* application, for significant problem sizes performed as well as CBC and Gurobi. With relaxation optimisation time was not a lynch-pin issue compared to the integrality of solutions.

Remarkably, through computational experiments, as the number of potential projects admitted to the portfolio grew, the number of non-integral solutions *did not*. This phenomenal result in practical terms meant that for large problem sets, the number of non-integral decision variables remained relatively small. Intuitively, the time structure of the multiple knapsacks (the yearly budgets) squeezes solutions that sit in two or more knapsacks, that is non-integral solutions out, forcing the project under consideration out of all knapsacks or into just one. This result, more than any other assured the practical viability of the NITRO application. Theoretically, other authors have observed and commented on this result in multiple knapsacks. From a very pragmatic point of view, detailed research is required into exactly what problem structure conditions are required for relaxation viability to be maintained.

Research into the optimisation of project portfolios will remain a fruitful activity for the foreseeable future. Without any specific order, the following problems must be addressed.

- Acceptance. The results of any portfolio optimisation must be sufficiently "close" to what an appropriate portfolio should be in the minds of the SMEs.
- The decision set. The simple binary decision to invest in a particular project (or not), at a specific time does not accurately reflect the decisions required for direct capability evolution. This abstraction gives us the advantage of drawing on the machinery of the knapsack problem. However an alternative is to develop an algorithmic allocation of yearly expenditure to a project itself. This optimisation process has a structure similar to that of the multi-period assignment problem. The disadvantage of this approach is one must specify yearly upper and lower bounds on project (not just budget) expenditure.
- Robustness of portfolios against variation in parameters. This issue is particularly pressing, in light of the semi-continuity of optimised values as parameters within the model change. There are now polynomial time approximation schemes (PTAS) for the single parametric knapsack problem [15], however a considerable amount of research is required in this area.
- Value. Our experience is that the value of a project is constantly being debated within the community of SMEs. In financial portfolio theory, by definition, there is a population-wide accepted consensus on value, that is money, however, that is not the case in this problem space. One approach is to avoid value altogether and instead to specify admissible project commencement times. This method has the disadvantage of optimisation infeasibility if there is no or constrained budget slack. Another approach is to specify a level of "temporal tolerance" in a project. Time-dependent values here are now reinterpreted as parameters that influence the timing of the project. An alternative approach is to consider solving a version of the inverse knapsack problem [44]. Recall that in the knapsack problem one decides to include a project in the portfolio based on its value, subject to costs. For the inverse knapsack problem one starts with the input of chosen projects and costs and finds an adjusted set of values $\mathbf{v}_{\text{new}}$ that minimises some norm $\|\mathbf{v}_{\text{new}} - \mathbf{v}_{\text{old}}\|$ with the old chosen value parameter set $\mathbf{v}_{\text{old}}$. Such norms are typically L_∞ or L_1.
- Uncertainty. There are multiple factors that drive uncertainty both across values and costs in Defence portfolio optimisation. As alluded to in the previous paragraph, there is currently no agreed definition of what exactly "value" is in the capability development context. Furthermore, even when value is chosen, for a project or group of projects, there is inherent uncertainty in this choice. Therefore, at the very least it would be useful for SMEs to choose from a range of values, perhaps drawn from a distribution (such as the maximum entropy distribution, which is uniform [25]). Algorithms, at least for the simplest form of the knapsack problem already exist [40].
- Project interactions. This issue has received a lot of attention in the literature, when the interactions between projects are important. One approach to project interactions is to assign an interaction score between projects and solve the so-called quadratic knapsack problem [26]. One could apply super-additive capacities in the interaction weights. However, rather than formulate as a quadratic knapsack problem, one could couple projects together in the time dependent knapsack problem through the use of temporal and precedence constraints. This has the advantage

of minimising the number of ambiguous values one needs to define, and takes advantage of the true problem structure.

Acknowledgements The Authors would like to thank Dr. Sharon Boswell, Dr. Yi Yue, and Dr. Lynn Booth for providing valuable feedback in the publication of this book chapter.

References

1. Archer, N., Ghasemzadeh, F.: A decision support system for project portfolio selection. Int. J. Technol. Manag. **16**(1/2/3), 105–114 (1998)
2. Archer, N.P., Ghasemzadeh, F.: An integrated framework for project portfolio selection. Int. J. Project Manage. **17**(4), 207–216 (1999)
3. Australia Research Council Centre for Excellence in Mathematical and Statistical Frontiers: The random sample-episode two, a not so secret linear programming paper, the story of Alison Harcourt (2019). https://acems.org.au/podcast/episode2-alison-harcourt. [Online; Accessed 12 Mar 2021]
4. Brown, T.: Design thinking. Harvard Business Review, pp. 84–93 (2008)
5. Burk, R.C., Parnell, G.S.: Portfolio decision analysis: lessons from military applications. Portfolio Decision Analysis: Improved Methods for Resource Allocation, pp. 333–357. Springer, Berlin (2011). https://doi.org/10.1007/978-1-4419-9943-6_14. http://discovery.ucl.ac.uk/1341061/
6. Chim, L., Nunes-Vaz, R., Prandolini, R.: Capability-based planning for Australia's national security. Secur. Chall. **6**(3), 79–96 (2010)
7. Dakin, R.J.: A tree-search algorithm for mixed integer programming problems. Comput. J. **8**(3), 250–255 (1965)
8. Davis, P.: Analytic architecture for capabilities-based planning, mission-system analysis, and transformation. Technical report, RAND Corporation (2002). https://apps.dtic.mil/sti/pdfs/ADA402243.pdf
9. Davis, P., Shover, R., Beck, J.: Portfolio-analysis methods for assessing capability options. Technical report, RAND Corporation (2008). https://www.rand.org/pubs/monographs/MG662.html
10. Department of Defence: The Defence Capability Manual (2020)
11. Forrest, J., Lougee-Heimer, R.: Cbc user guide. INFORMS TutORials in Operations Research, pp. 257–277 (2019). https://pubsonline.informs.org/doi/pdf/10.1287/educ.1053.0020
12. Forrest, J., Robin, L.H.: CBC User Guide. https://www.coin-or.org/Cbc/cbcuserguide.html#id3413036
13. Frederick, S., Loewenstein, G., O'Donoghue, T.: Time discounting and time preference: a critical review. J. Econ. Lit. **40**, 351–401 (2002)
14. Gass, S.: Linear Programming Methods and Applications. McGraw-Hill, New York (1969)
15. Giudici, A., Halffmann, P., Ruzika, S., Thielen, C.: Approximation schemes for the parametric knapsack problem. Inf. Process. Lett. **120**, 11–15 (2017)
16. Goldberg, A., Marchetti-Spaccamela, A.: On finding the exact solution of the zero-one knapsack problem. In: Proceedings of the 16th Annual ACM Symposium on the Theory of Computing. ACM (1984)
17. Hadigheh, A., Terlaky, T.: Sensitivity analysis in linear optimisation: invariant support set intervals. Optim. Online (2004). http://www.optimization-online.org/DB_FILE/2004/08/928.pdf
18. Hansen, J., Lipow, J.: Time discounting in military cost benefit analysis. Military Cost-Benefit Analysis: Theory and Practice, pp. 424–431. Routledge, Oxfordshire (2015)
19. Harrison, K.R.: Portfolio costs and values dataset (2019). Data from the University Of New South Wales School of Information Technology and Engineering

20. Harrison, K.R., Elsayed, S., Garanovich, I., Weir, T., Gallister, M., Boswell, S., Taylor, R., Sarker, R.: Multi-period project selection and scheduling for defence capability-based planning. In: 2020 IEEE International Conference on Systems, Man, and Cybernetics, pp. 4044–4050. IEEE (2020)
21. Harrison, K.R., Elsayed, S., Garanovich, I., Weir, T., Gallister, M., Boswell, S., Taylor, R., Sarker, R.: Portfolio optimization for defence applications. IEEE Access **8**(1), 60152–60178 (2020)
22. Hayes, B.: Relationship between force design and innovation. UNSW ADFA (2017). https://www.unsw.adfa.edu.au/conferences/sites/conferences/files/uploads/user50/ASDCI-Presentation-Force-Design-Conference-10oct2017.pdf
23. Heidenberger, K., Stummer, C.: Research and development project selection and resource allocation: a review of quantitative modelling approachess. Int. J. Manag. Rev. **1**, 197–224 (1999)
24. Hupfeld, M.: MilCIS 2017 - Head Force Design Presentation. University of New South Wales (2017). https://static1.squarespace.com/static/527411 2ae4b02d3f058d4348/t/5a14ef6fe2c483f5ad33e530/1511321459977/2017-1-3a.pdf
25. Jaynes, E.T.: On the rationale of maximum-entropy methods. Proc. IEEE **70**(9), 939–952 (1982)
26. Kellerer, H., Pferschy, U., Pisinger, D.: Knapsack Problems. Springer, Berlin (2004). https://doi.org/10.1007/978-3-540-24777-7. https://www.springer.com/gp/book/9783540402862
27. Koc, A., Morton, D.P., Popova, E., Hess, S.M., Kee, E., Richards, D.: Prioritizing project selection. Eng. Econ. **54**(4), 267–297 (2009). Taylor & Francis. https://doi.org/10.1080/00137910903338545
28. Kronqvist, J., Bernal, D.E., Lundell, A., Grossmann, I.: A review and comparison of solvers for MINLP. Eng. Optim. **20**, 397–455 (2019). URL https://link.springer.com/article/10.1007
29. Laibson, D.: Golden eggs and hyperbolic discounting. Q. J. Econ. **112**(2), 443–477 (1997). https://dash.harvard.edu/bitstream/handle/1/4481499/Laibson_GoldenEggs.pdf
30. Land, A.H., Doig, A.G.: An automatic method of solving discrete programming problems. Econometrica **28**(3), 497–520 (1960)
31. Lin, Y., Schrage, L.: The global solver in the LINDO API. Optim. Methods Softw. **24**, 657–668 (2009)
32. Markowitz, H.: Portfolio selection. J. Financ. **7**, 77–91 (1952)
33. Martello, S., Toth, P.: Knapsack Problems Algorithms and Computer Implementations. Wiley, Hoboken (1990)
34. Martin, D.: On the continuity of the maximum in parametric linear programming. J. Optim. Theory Appl. **17**, 205–2010 (1975). https://link.springer.com/content/pdf/10.1007/BF00933875.pdf
35. Marzec, M.: Portfolio optimization: applications in quantum computing. Handbook of High-Frequency Trading and Modeling in Finance, pp. 73–106 (2016)
36. McGinty, N.: Force Exploration & DST. UNSW ADFA (2017). https://www.unsw.adfa.edu.au/conferences/sites/conferences/files/uploads/user50/Force-Design-Confernce-Presentation-10oct2017.pdf
37. McKenna, T., McKay, T.: Australia's joint approach: past, present and future. Joint Studies Paper Series No, 1. Canberra: Defence Publishing Service (2017). https://www.defence.gov.au/ADC/Publications/documents/joint_studies/JSPS_1_Australias_Joint_Approach.PDF
38. Merkle, R., Hellman, M.: Hiding information and signatures in trapdoor knapsacks. IEEE Trans. Inf. Theory **24**, 525–530 (1978)
39. Mohagheghi, V., Mousavi S.M., Antuchevičienė, J., Mojtahedi, M.: Project portfolio selection problems: a review of models, uncertainty approaches, solution techniques, and case studies. Technol. Econ. Dev. Econ. **25**(6), 1380–1412 (2019)
40. Morton, D., Wood, R.: On a stochastic knapsack problem and generalisations. In: Woodruff, D. (ed.) Advances in Stochastic Optimisation, Logic Programming and Heuristic Search, 9, vol. 1, chap. 5, pp. 149–168. Springer Science (1998)
41. Peacock, J., Blumson, D., Mangalasinghe, J., Hepworth, A.C., Lo, E.H.S.: Baselining the whole-of-force capability and capacity of the Australian defence force. In: Elsawah, S. (ed.)

MODSIM2019, 23rd International Congress on Modelling and Simulation. Modelling and Simulation Society of Australia and New Zealand, December, pp. 249–255 (2019). https://doi.org/10.36334/modsim.2019.B3.peacock
42. Ponsich, A., Jaimes, A.L., Coello, C.A.C.: A survey on multiobjective evolutionary algorithms for the solution of the portfolio optimization problem and other finance and economics applications. IEEE Trans. Evol. Comput. **17**(3), 321–344 (2013)
43. Puchinger, J., Raidl, G., Pferschy, U.: The multidimensional knapsack problem: structure and algorithms. INFORMS J. Comput., Inst. Oper. Res. Manag. Sci. (INFORMS) **22**(2), 250–265 (2010)
44. Rolanda, J., Figueira, J.R., De Smet, Y.: The inverse 0–1 knapsack problem: theory algorithms and computational experiments. Discret. Optim. **10**, 181–192 (2017)
45. Samuelson, P.: A note on measurement of utility. Rev. Econ. Stud. **4**, 155–161 (1937)
46. Spiegeleire, S.D., van Hooft, P., Culpepper, C., Willems, R.: Closing the Loop: Towards Strategic Defence Management. The Hague Centre for Strategic Studies (2009)
47. Stanley, E.D., Honig, D.P., Gainen, L.: Linear programming in bid estimation. Nav. Res. Logist. Q. **1**, 48–54 (1954)
48. Tate, D.M., Thompson, P.M.: Portfolio Selection and Resource Allocation for Defense Applications. Institute for Defense Analysis (2016)
49. Tate, D.M., Thompson, P.M.: Portfolio Selection Challenges in Defense Applications. Institute for Defense Analysis (2017)
50. Cambridge University Dictionary: "portfolio" (2020). https://dictionary.cambridge.org/dictionary/english/portfolio
51. Commonwealth of Australia: Integrated Investment Program. Commonwealth of Australia Department of Defence (2016)
52. Commonwealth of Australia: 2020 Defence Strategic Update. Commonwealth of Australia Department of Defence (2020). https://www1.defence.gov.au/sites/default/files/2020-11/2020_Force_Structure_Plan.pdf
53. Commonwealth of Australia: 2020 Force Structure Plan. Commonwealth of Australia Department of Defence (2020). https://www1.defence.gov.au/sites/default/files/2020-11/2020_Force_Structure_Plan.pdf
54. Department of Defence: The Strategy Framework, 2017. Australian Government Department of Defence (2017). https://www.defence.gov.au/SPI/Docs/The
55. El Hadi, M.: Towards capabilities-based planning to enhance and optimize sustainable development within proper enterprise architecture, vol. 26, pp. 1–30. Scientific Conference of Information Systems & Computer Technology (2018)
56. The Department of Defence: The Auditor-General ANAO Report No. 26. 2016–17 Major Projects Report. Commonwealth of Australia Department of Defence (2017). https://www.anao.gov.au/work/major-projects-report/major-projects-report-2016-17
57. The Technical Cooperation Program: Guide to Capability Based Planning. US/UK/Canada/Australia/NZ Technical Cooperation Program (2004)
58. Thie, P.R., Keough, G.E.: An Introduction to Linear Programming and Game Theory, 3rd edn. Wiley, Hoboken (2008)
59. Wall, K.D., MacKenzie, C.A.: Multiple objective decision making. Military Cost-Benefit Analysis: Theory and Practice, pp. 197–236. Routledge, Oxfordshire (2015)
60. Weir, T.: Regression models for project expenditures. In: Sarker, R., Abbass, H.A., Dunstall, S., Kilby, P., Davis, R., Young, L. (eds.) Data and Decision Sciences in Action. Lecture Notes in Management and Industrial Engineering, pp. 271–290. Springer International Publishing, Cham (2018). https://doi.org/10.1007/978-3-319-55914-8_20
61. Weir, T., Johns, K.: Longitudinal models for project expenditure plans. In: Proceedings of the 22nd International Congress on Modelling and Simulation, December, pp. 695–701 (2017). https://doi.org/10.36334/MODSIM.2017.D3.Weir. http://www.mssanz.org.au/modsim2017/D3/weir.pdf
62. Wijnmalen, D., Curtis, N.: A code of best practice for judgement-based operational research. OR Insight **26**(4), 291–308 (2013)

63. Wilbaut, C., Hanafi, S., Salhi, S.: A survey of effective heuristics and their application to a variety of knapsack problems. IMA J. Manag. Math. **19**, 227–244 (2007)
64. Williams, T., Vo, H., Samset, K., Edkins, A.: The front-end of projects: a systematic literature review and structuring. Prod. Plan. Control. **30**(14), 1137–1169 (2019)
65. Wrigley, C., Mosely, G., Gill, M., Reid, J.: Air force by design: applying design for transient capability advantages. Joint Studies Paper Series No, 3. Canberra: Defence Publishing Service (2019). https://www.defence.gov.au/ADC/Publications/documents/joint_studies/JSPS_3_Design_Thinking.pdf
66. Zwikael, O., Chih, Y., Meredith, J.: Project benefit management: setting effective target benefits. Int. J. Project Manage. **36**, 650–658 (2018)
67. Zwikael, O., Garanovich, I.L., Weir, T., Gowlett, P., Aljoofri, S., Ween, A.: Identifying triggers for new projects and programs at the Australian department of defence. In: Proceedings of the Australian Institute of Project Management National Conference, pp. 27–38 (2018)
68. Zwikael, O., Gilchrist, A.: The logic of the front-end. In: Williams, T., Samset, K., Volden, G.H. (eds.) Paradoxes and Ways Ahead: On Front-End Management of Projects (2021)
69. Zwikael, O., Meredith, J.: Effective organizational support practices for setting target benefits in the project front end. Int. J. Project Manage. **37**(7), 930–939 (2019)
70. Zwikael, O., Smyrk, J.R.: Project Management for the Creation of Organisation Value. Springer, Berlin (2011)
71. Zwikael, O., Smyrk, J.R.: Project Management: A Benefit Realisation Approach. Springer, Berlin (2019)

A Decision Support System for Planning Portfolios of Supply Chain Improvement Projects in the Semiconductor Industry

Rainer Kolisch and Thomas Fliedner

Abstract We report on a decision support system to select supply chain improvement projects of an international semiconductor manufacturer. At the core of the system is an integer programming model optimizing the value of the project portfolio while taking into account project interdependencies and multiple objectives via aspiration levels. We present the graphical user interface of the decision support system and provide insight on the latter via a case study. Furthermore, we report on the feedback of the decision makers using the system.

1 Introduction

In this chapter we present a decision support system (DSS) to select supply chain improvement projects of an international semiconductor manufacturer. While the company has established a clear process for project evaluation and prioritizing overseen by a project portfolio committee, company decision makers have expressed difficulties in assessing project proposals evaluated in terms of multiple criteria and subject to interdependencies. When considered in isolation, a project proposal with high cost and low value may be rejected by traditional evaluation techniques, e.g., the value-for-money principle [15, 22, 28]. Nonetheless, the project may be valuable to the portfolio as a whole as it enables the execution of other valuable projects or is rendered profitable due to synergistic effects in combination with other projects. De Reyck et al. [4] emphasize adequate consideration of project interdependencies, incorporation of selection constraints, as well as proper alignment of the portfolio to company strategy as key elements of project portfolio management and drivers of project success. To this end, a decision support system has been developed to provide

R. Kolisch (✉)
Technical University of Munich, Arcisstr. 21, 80333 München, Germany
e-mail: rainer.kolisch@tum.de

T. Fliedner
A.T. Kearney, Sendlinger Str. 8, 80331 München, Germany
e-mail: thomas.fliedner@kearney.com

K. R. Harrison et al., *Evolutionary and Memetic Computing for Project Portfolio Selection and Scheduling*, Adaptation, Learning, and Optimization 26,
https://doi.org/10.1007/978-3-030-88315-7_8

guidance in the decision making process. The system was designed to provide decision makers with portfolio recommendations that are aligned with the supply chain strategy and provide highest possible financial contribution. The system considers project interdependencies and multiple objectives. At the core of the system is an integer programming model taking into account logical interdependencies between projects as well as synergies. Strategic alignment of the portfolio recommendation is ensured by allowing decision makers to express aspiration levels for evaluation criteria. A graphical user interface enables the users to visualize the obtained portfolio and its performance with respect to the different objectives. Furthermore, users can overule optimal solutions by forcing projects in the portfolio or out of the project portfolio. We illustrate the proposed decision support system and its underlying integer program by presenting a case study on the selection of supply chain improvement projects of a semiconductor manufacturer. Finally, we provide feedback from the decision makers about using the decision support system.

The remainder of this chapter is organized as follows. In Sect. 2 we review the most relevant literature and position our work. Section 3 proposes the formal decision making framework and the integer programming model for project portfolio selection, which has been embedded in a decision support system introduced in Sect. 4. We demonstrate the proposed model in a case study in Sect. 5 before providing conclusions and an outlook to future research in Sect. 6.

2 Literature

There is a rich literature on project porfolio planning. Reviews are given by, e.g., [1, 12, 13], and recently by [25]. Morton et al. [26] provide a broad treatment of project portfolio planning from the perspective of multi-criteria portfolio decision analysis. In what follows, we review a selection of the work in the project portfolio domain. Thereby, we particular consider project interdependencies, multiple objectives, and decision support systems.

Santhanam and Kyparisis [32] propose an integer programming model for information system project selection taking into account logical interdependencies as well as synergies between projects. They solve a real-world instance from a service organization where information system projects have to be selected.

Stummer and Heidenberger [35] develop a multi-objective integer programming model for the selection of research and development projects. All Pareto optimal project portfolios are determined through complete enumeration, which the authors regard as computationally tractable for instances with up to 30 projects. The model takes into account logical interdependencies and synergies between projects as well as aspiration levels for the relevant criteria.

Kolisch et al. [18] present an integer programming model for solving the project portfolio planning problem, which builds up on the works of [32, 35]. They take into account multiple criteria and project interdependencies. The model is embedded into a decision support system and applied to a real-world problem of selecting a portfolio of supply chain improvement projects.

Dickinson et al. [5] provide decision support for the selection and timing of technology projects at Boeing. A non-linear integer programming model is used to maximize the net present value of the portfolio subject to a budget constraint as well as aspiration levels on the strategic alignment of projects in the portfolio. Logical interdependencies between projects are modeled with a dependency matrix, which is also used to transfer value between interdependent project proposals.

Grushka-Cockayne et al. [10] provide a decision making framework for the selection of operational improvement projects and their execution modes at the European air traffic management organization. The authors consider a linear-additive portfolio value model taking into account project scores evaluated by multiple stakeholders according to multiple criteria. An integer programming model is developed that maximizes overall portfolio value while taking into account multiple budget restrictions and aspiration levels for all criteria. The model is extended to additionally account for non-linear project synergies by considering decisions on project clusters instead of single projects.

Gurgur and Morley [11] support the selection of maintenance and infrastructure projects at the Lockheed Martin Space Systems Company. A multi-criteria linear-additive portfolio value model is developed, where criteria weights are determined by swing-weighting. The authors propose an integer programming model that maximizes aggregate project utility. Uncertainty of the cost and the duration of projects is taken into account by chance-constraints limiting the probability of time and cost overruns.

Chien and Huynh [2] consider the problem of selecting and scheduling integrated circuit design R&D-projects of semiconductor companies. They propose an optimization model with multiple objectives as well as synergies between projects. For solving the model, they develop a genetic algorithm. They apply their approach to real-world instances of an integrated circuit design company.

Fliedner and Liesiö [7] address the multi-attribute portfolio selection problem where the overall objective function is a weighted sum of the attributes of selected projects with incomplete information of the weights. Furthermore, the attributes of projects are uncertain within given ranges. Using adjustable robust optimization, the authors propose an approach to determine non-dominated portfolios. They apply their approach to data of supply chain improvement projects of a semiconductor manufacturer.

Hu and Szmerekovsky [14] treat the problem of selecting projects and assigning a budget to each selected project, where the cost of each project is modeled with a distribution function. The authors apply a newsvendor approach, which minimizes the expected profit of the portfolio minus the sum of the overage and underage costs of the selected projects. Overage cost of a project accrues if its cost is above the assigned budget, while underage cost accrues when the cost of a selected project is below its assigned budget.

Panadero et al. [27] consider a stochastic project portfolio problem, where the size of selected projects can be varied within given limits. The cash flow of generated projects is uncertain and the objective is to maximize the expected net present value of the portfolio. For solving the problem, an adaptive variable neighborhood heuristic is proposed, which employs a Monte Carlo Simulation. Experimental results are obtained for artificially generated instances.

Roland et al. [29] consider a project portfolio planning problem with multiple experts, each having a value for each project. The expert's value for a project portfolio is the sum of the expert's values of the selected projects. Experts have individual weights, which sum up to 1 for all experts. The objective is to determine the portfolio that minimizes the weighted l_p-norm of the distances of the selected portfolio to the optimal portfolio of each expert. The problem is cast as a multi-objective problem and solved with an optimal and a heuristic cutting plane approach. Experimental results on artificially generated instances are undertaken.

Sampath et al. [30] present a decision support framework for project portfolio planning at Intel Corporation. The authors consider a bi-criteria objective function, different types of project interdependencies, and propose an integer programming model. Since the latter is encountering running time problems when solving large real-world instances, Sampath et al. [31] extend this work by proposing an evolutionary algorithm.

Undertaking behavioral experiments, Schiffels et al. [33] investigate how human decision makers determine project portfolios.

Stewart [34] addresses the project portfolio problem with project interdependencies and multiple objectives. Interdependencies are modeled by generating so-called "metaprojects", artificial projects which include individual projects connected via interdependencies. A genetic algorithm is proposed to generate the efficient frontier of project portfolios.

Against the background of the literature, our work is positioned as follows. As in [2, 10, 30, 31], our work is initiated by and addressing a practical application. As in [10, 30, 31, 34], our project portfolio planning problem is taking into account interdependencies as well as synergies between projects. Furthermore, we are considering multiple objectives, as in [7, 29–31, 34]. We are doing so by using an aspiration level approach, see [35]. We are taking into account risk, as in [27]. This is done in a straightforward way by using a project specific risk parameter by which project scores are multiplied. The resulting project portfolio selection problem is modeled and solved to optimality by employing an integer programming approach, as it is done in [30]. By contrast to [30], we are proposing a less demanding integer programming model, which in turn does not lead to running time problems for large instances experienced by [30]. Finally, we are embedding our solution approach into a decision support system, as done by [5, 18, 30, 31, 35].

Our work is closest to [30, 31] but differs from the latter in a number of ways. First, we model synergies in a less demanding way such that all instances can be solved to optimality with a commercial solver. Second, we don't generate efficient frontiers but use aspiration criteria. Third, we consider more than two objectives.

3 Decision Making Framework and Integer Programming Model

Initially, project owners develop business cases for their project proposals. Using a standardized template, they specify qualitative objectives, the scope, as well as the deliverables of their proposal. Formally, projects $j = 1, \ldots, m$ are evaluated in terms of execution cost giving a cost vector $c \in \mathbb{R}^m_+$ and multiple score criteria $i = 1, \ldots, n$ giving a score matrix $v \in \mathbb{R}^{m \times n}_+$. The additive overall value of project j is $\sum_{i=1}^{n} w_i v_{ji}$, where the vector of criterion weights $w \in \mathbb{R}^n_+$ relates a unit increase in the criterion-specific score to an increase in the overall project value. Specifically, the potential of supply chain improvement through projects is evaluated in terms of their financial and strategic value. Projects can create financial value by reducing cost within the company supply chain or by enabling additional revenue for the company. Strategically, projects can improve the lead time of customer orders, termed "Speed", or the accuracy of forecasted customer demands, called "Forecast Accuracy" (F/A). While financial value is measured in monetary units, the Speed criterion is measured in days and Forecast Accuracy in percentage points. Weights to translate strategic scores to monetary units have been determined by a previous internal study on supply chain strategy.

Project portfolios are subject to multiple sources of uncertainty stemming from the company environment, project interdependencies, or the projects themselves [23]. Different approaches of tackling uncertainty in project selection problems have been proposed: The specification of distributions for uncertain data [11], the definition of discrete scenarios of parameter realizations [20], the interval representation of input data [6], as well as the representation of uncertain information by fuzzy numbers [36]. The process established at the semiconductor manufacturer requires project owners to specify a risk parameter with which nominal project scores are multiplied to obtain risk-adjusted scores v. This approach is in line with common practices of project risk management [3, 28], yet follows the paradigm of interpreting uncertainty primarily as a threat to portfolio success [37].

Evaluated project proposals are discussed, pre-screened, and prioritized during monthly core team meetings by representatives from different departments. At this stage, interdependencies between project proposals are identified. Firstly, projects can be logically connected, i.e., the execution of one project is dependent on the execution of one or more other projects. As an example, a project to improve company revenue by implementing decision support for supplier negotiations may be dependent on the execution of an independent project to introduce a database on previous procurement contracts. Secondly, synergies may arise from the execution of two or more related projects. Greater cost efficiency of joint project execution creates additional financial value, whereas greater effectiveness in supporting the supply chain strategy creates additional value in terms of Speed or Forecast Accuracy. During quarterly steering committee meetings, pre-screened project proposals are presented and a portfolio is selected for execution subject to budget restrictions.

In order to aid decision makers during steering committee meetings, portfolio recommendations are developed by solving an integer programming problem for project portfolio selection. A project portfolio is modeled as a binary vector $x \in \{0, 1\}^m$ indicating the selection ($x_j = 1$) of project $j = 1, \ldots, m$. The overall portfolio value is given by the sum of the additive overall values of selected projects [9]. Synergies $s = 1, \ldots, o$ account for non-additive beneficial effects $v^S \in \mathbb{R}_+^{o \times n}$ resulting from the joint execution of related projects [32]. They are modeled as a binary vector $y \in \{0, 1\}^o$ indicating the application ($y_s = 1$) of synergy $s = 1, \ldots, o$ if all required projects $j \in \mathcal{J}_s$ are selected.

The integer programming problem solved in the decision support system is given by

$$\text{maximize} \sum_{i=1}^{n} \sum_{j=1}^{m} v_{ji} w_i x_j + \sum_{i=1}^{n} \sum_{s=1}^{o} v_{si}^S w_i y_s - \sum_{j=1}^{m} c_j x_j \tag{1}$$

subject to the constraints

$$\sum_{j=1}^{m} v_{ji} x_j + \sum_{s=1}^{o} v_{si}^S y_s \geq v_i^{\text{asp}} \qquad \forall i = 1, \ldots, n \tag{2}$$

$$\sum_{j=1}^{m} c_j x_j \leq B \tag{3}$$

$$x_{j'} \leq x_j \qquad \forall (j, j') \in \mathcal{E} \tag{4}$$

$$\sum_{j \in \mathcal{J}_s} x_j \geq |\mathcal{J}_s| y_s \qquad \forall s = 1, \ldots, o \tag{5}$$

$$x_j, y_s \in \{0, 1\} \qquad \forall j = 1, \ldots, m \quad s = 1, \ldots, o. \tag{6}$$

By company policy, projects may only receive funding if they are profitable, i.e., their financial value is higher than their cost, or if they are essential in order to achieve decision makers' strategic goals for the company supply chain. To this end, the integer programming problem is solved to recommend project portfolios that provide the highest possible value contribution (1), i.e., the sum of the portfolio value and the aggregate value of synergies reduced by the aggregate cost of selected projects. At the same time, aspiration level constraints (2) ensure minimum achievement levels $v^{\text{asp}} \in \mathbb{R}_+^n$ for each score criterion [17, 35], particularly reflecting strategic goals in terms of Speed and Forecast Accuracy. The budget constraint (3) specifies that the aggregate cost of all executed projects may not exceed a given budget level $B \in \mathbb{R}_+$. Logical interdependencies between projects are modeled by constraints (4). The set $\mathcal{E}$ contains all project pairs $(j, j') \in \mathcal{E}$ where project j' can only be selected if another project j required for its execution is selected as well [32]. Finally, constraints (5) ensure that synergy values are only taken into account if the required projects are selected.

4 Decision Support System

Marx et al. [24] have performed an empirical study to derive six principles for designing successful decision support systems. They emphasize system ease-of-use, system flexibility and adaptability, as well as a comprehensive and well-manageable information model. A trade-off arises between the comprehensiveness of information and support system usability. Similar aspects were emphasized by decision makers within the semiconductor manufacturer during development of the decision support system.

To aid usability of the decision support system, relevant project and portfolio information should be displayed in a clear fashion and the system should employ visualizations and wording aligned to the established project selection process. Decision support system interactivity was of prime importance to the decision makers. Displayed information and portfolio recommendations should be easily adjustable during meetings and should support live adjustments to aspiration levels and manual selection or exclusion of projects. Furthermore, to ensure information consistency and minimize system maintenance, the decision support system, which requires information on projects, project interdependencies, as well as the project portfolio environment, should seamlessly integrate with existing data sources.

Taking into account these requirements, a decision support system had been developed embedding the integer programming model presented in Sect. 3 to aid the project portfolio selection process during steering committee meetings.

The decision support system user interface consists of an interactive browser client integrated in the internal network of the company. The back-end system builds and solves the integer programming problem and manages the presentation of project and portfolio information in the interface.

Figure 1 provides an illustration of the user interface. At the bottom left of the interface, a list of all project proposals is displayed (Section "Project List"). The list contains the name of each project as well as its financial value. By clicking on any project, a window with detailed information appears. In the top right part of the interface, slide bars allow decision makers to adjust aspiration levels as a percentage of maximum attainable criteria scores $v^{\max} \in \mathbb{R}^n_+$. When the decision support system is started, maximum scores for each criterion $i = 1, \ldots, n$ are determined by solving the integer programming problem

$$v_i^{\max} = \text{maximize} \sum_{j=1}^{m} v_{ji} x_j + \sum_{s=1}^{o} v_{si}^S y_s \tag{7}$$

subject to constraints (3)–(6) for each score criterion. Absolute aspiration levels are given below the slide bars and are automatically adjusted whenever the slide bars are changed.

By clicking the "Solve" button, decision makers can invoke the decision support system to solve integer programming problem (1)–(6) taking into account the chosen aspiration levels v^{asp}. The resulting portfolio recommendation, i.e., a list of all

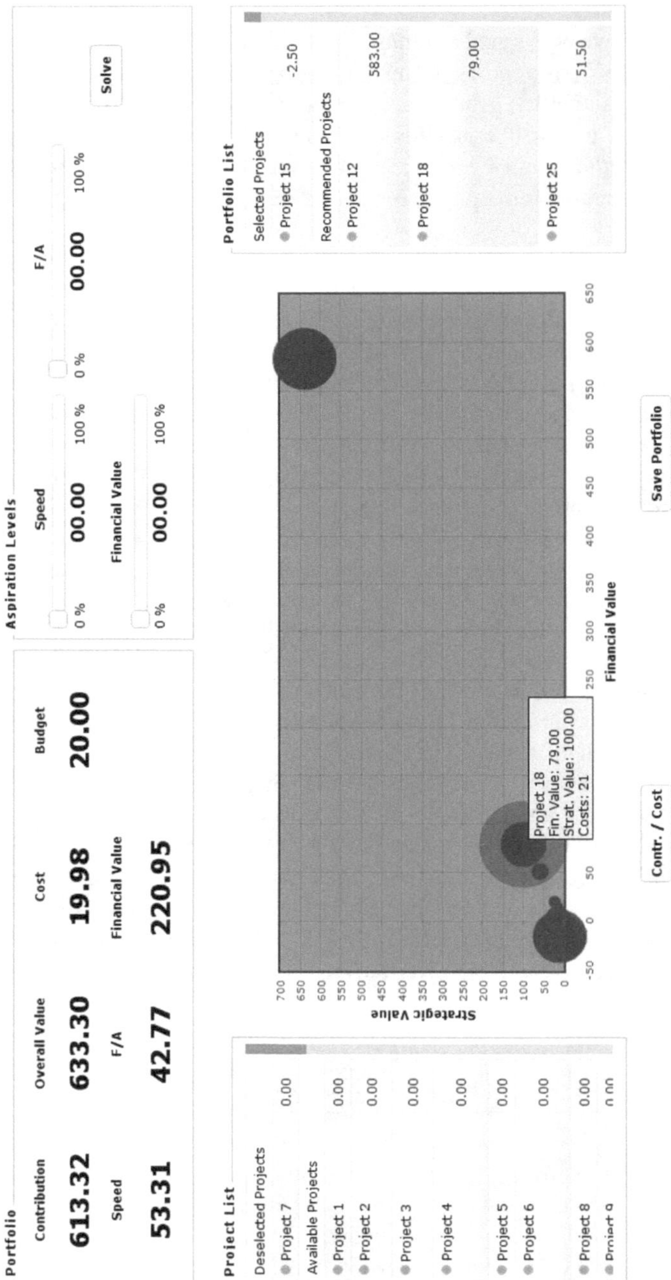

Fig. 1 Illustration of the decision support system interface

selected projects, is displayed at the bottom right of the interface (Section "Portfolio List"). In the center of the interface, a bubble chart [3] of the recommended portfolio is displayed. All selected projects are plotted in the space of their financial and strategic impact, i.e., the sum of projects' monetized Speed and Forecast Accuracy scores. Logical interdependencies of pairs of projects can be visualized in a dependency matrix [5, 16], a matrix with all projects in the rows and in the columns. A matrix-element of 1 in row j and column j' indicates that project j' can only be selected if project j is selected.

The top left "Portfolio" Section contains information on the budget limit as well as aggregate information on the recommended project portfolio. Portfolio cost, aggregate values for each score criterion as well as the overall portfolio value are presented. Furthermore, the section presents the objective function value of the integer programming problem (1)–(6), the portfolio contribution.

Decision makers can manually select projects by dragging them from the "Project List" to the "Portfolio List". Projects may be excluded from the portfolio by dragging them to the edge of the decision support system screen. Whenever a project is manually selected or excluded, the integer programming problem (1)–(6) is automatically solved again taking into account decision makers' manual adjustments. If the integer programming problem is infeasible, a notification is displayed prompting decision makers to adjust their choices.

5 Case Study

We report computational results obtained during a past iteration of the project selection process. Table 1 summarizes all relevant project information, see also [18]. 58 projects (named P1 to P58) are evaluated in terms of $n = 3$ criteria: Speed ($i = 1$), Forecast Accuracy ($i = 2$), and financial value ($i = 3$). While financial value is directly measured in monetary units ($w_3 = 1$), improving Speed by one unit is worth $w_1 = 2.6$ monetary units, and one unit improvement in Forecast Accuracy is worth $w_2 = 6.4$ monetary units. 13 projects can only be selected if other projects are selected as well. Overall, 26 logical interdependencies between projects exist.

Synergies are summarized in Table 2. For example, the joint selection of projects P9 and P27 creates additional value $v^S_{1,3} = 5.46$, amounting to 19% of the overall value of the two projects (synergy S1).

We report optimal solutions to integer programming problem (1)–(6) for different budget levels B and no restrictions on aspiration levels ($v_1^{\text{asp}} = v_2^{\text{asp}} = v_3^{\text{asp}} = 0$). Overall, the integer programming problem contains 61 decision variables and 33 constraints. Optimal portfolios for all budget levels $B = 5, 10, 15, \ldots, 125$ were determined in less than five seconds (IBM ILOG CPLEX 12.6, 3.2 GHz dual-core processor, 8 GB memory). For most budget levels, unique optimal solutions exist. Only for budget levels $B = 25$ and $B = 60$, ten distinct optimal portfolios exist.

Figure 2 reports the value composition of optimal portfolios. Portfolio value with regards to each score criterion $i = 1, \ldots, n$ is given by $\sum_{j=1}^{m} v_{ji} w_i x_j$. Portfolio value

Table 1 Project scores, cost, and logical interdependencies

j	v_{j1}	v_{j2}	v_{j3}	c_j	Required projects
	$w_1 = 2.6$	$w_2 = 6.4$	$w_3 = 1$		
P1	0.0	0.0	0.4	0.44	
P2	0.0	35.57	0.0	1.24	
P3	0.0	0.0	29.38	1.2	
P4	0.0	0.0	28.32	0.52	
P5	0.0	0.0	29.68	16.2	
P6	0.0	0.0	112.42	51.82	P7, P8, P9, P10, P11, P28
P7	0.0	0.67	0.0	0.62	
P8	2.0	0.0	0.0	0.22	
P9	0.0	0.0	1.02	0.26	
P10	0.0	7.2	0.0	1.4	
P11	0.0	0.67	0.0	0.96	
P12	0.0	0.0	48.96	4.98	P27, P33, P34
P13	0.0	0.0	8.63	6.84	
P14	0.0	0.0	0.33	0.61	
P15	0.0	0.0	0.35	0.62	
P16	0.0	0.0	7.32	0.22	P37
P17	0.0	0.0	0.45	0.62	
P18	0.0	0.0	21.95	1.0	
P19	5.99	0.0	0.0	0.34	
P20	13.64	0.0	0.0	0.78	
P21	1.66	0.0	0.0	0.22	
P22	4.09	0.0	0.0	0.88	P8
P23	3.29	0.0	0.0	0.58	P19, P20, P21, P22, P24, P25, P26
P24	1.64	0.0	0.0	0.34	P8
P25	13.64	0.0	0.0	1.26	
P26	5.46	0.0	0.0	1.34	P27
P27	0.0	0.0	27.29	2.48	P28
P28	0.0	0.0	27.29	1.7	P29
P29	0.0	0.0	17.56	2.3	
P30	0.0	0.0	2.73	0.38	
P31	0.0	0.71	2.11	1.18	P34
P32	0.0	0.0	29.33	2.14	
P33	0.0	0.0	18.19	12.86	
P34	0.0	0.0	1.23	0.64	P37

(continued)

Table 1 (continued)

j	v_{j1}	v_{j2}	v_{j3}	c_j	Required projects
	$w_1 = 2.6$	$w_2 = 6.4$	$w_3 = 1$		
P35	0.0	0.0	0.14	0.74	P31
P36	0.0	0.0	0.67	0.22	
P37	6.55	0.0	1.64	0.34	
P38	0.0	0.0	0.8	0.22	
P39	0.0	0.0	1.06	0.22	
P40	0.0	0.0	1.72	0.22	
P41	0.0	0.0	1.66	0.22	P37
P42	0.0	0.0	8.01	0.22	
P43	0.0	0.0	2.77	0.22	
P44	0.0	0.0	3.33	1.34	
P45	0.0	0.0	0.67	0.22	
P46	0.0	0.0	6.65	1.6	
P47	0.0	0.0	0.33	0.12	
P48	0.0	0.0	1.0	0.1	
P49	0.0	0.0	0.33	0.08	
P50	0.0	0.0	0.33	0.16	
P51	0.0	0.0	0.4	0.1	
P52	0.0	0.0	0.4	0.05	
P53	0.0	0.0	0.4	0.1	
P54	0.0	0.0	0.4	0.1	
P55	0.0	0.0	0.4	0.2	
P56	0.0	0.0	0.4	0.38	
P57	0.0	0.0	0.4	0.1	
P58	0.0	0.0	0.4	0.1	

Table 2 Synergy scores

s	v^S_{s1}	v^S_{s2}	v^S_{s3}	Required projects
	$w_1 = 2.6$	$w_2 = 6.4$	$w_3 = 1$	
S1	0.0	0.0	5.46	P9, P27
S2	4.09	0.0	8.19	P20, P27
S3	0.0	0.0	0.74	P11, P31

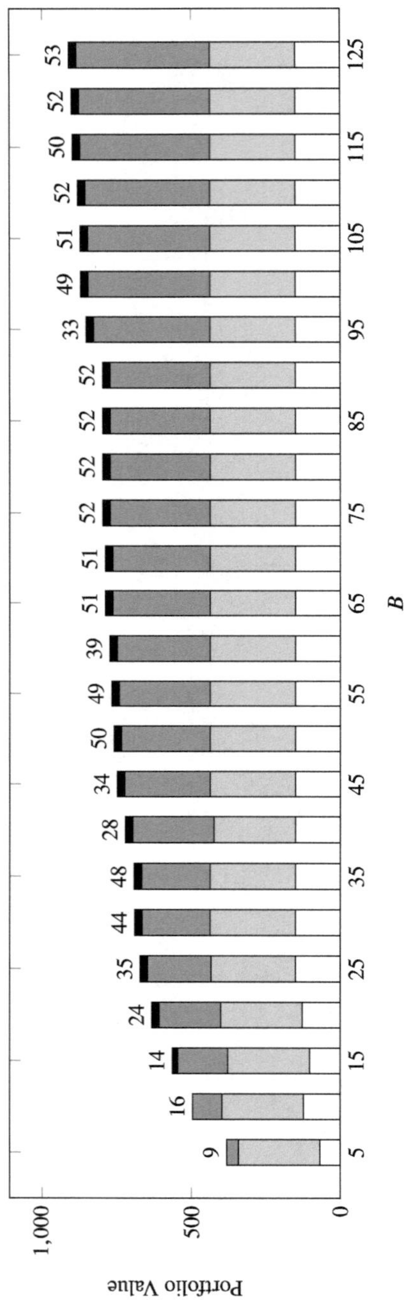

Fig. 2 Value composition of optimal portfolios for given budget levels

due to synergies is calculated as $\sum_{i=1}^{n}\sum_{s=1}^{o} v_{si}^{S} w_i y_s$. The number of selected projects $\sum_{j=1}^{m} x_j$ is shown above the bar charts. For low budget levels $B = 5, \ldots, 25$, we find that portfolios are dominated by projects providing strategic impact in terms of Speed and particularly Forecast Accuracy. Scores in strategic criteria initially rise quickly and reach their peak at budget level $B = 30$, whereafter additional portfolio value is primarily driven by financial value. Value from synergies is utilized from budget level $B = 15$ onward. Value contribution from Speed is reduced when increasing the budget from $B = 10$ to $B = 15$ in order to accommodate projects that provide higher financial value.

Figure 3 reports the relative frequency of each project and each synergy being selected in an optimal portfolio for varying budget levels. While relative frequencies are either zero or one when there is only one unique optimal solution, fractional values are possible for $B = 25$ and $B = 60$ where multiple optimal solutions exist. High relative frequencies of selection across varying budget levels give an indication of the value of a project to decision makers. A similar approach termed "core index" is employed by [19]. We find that projects P2, P4, P10, P19, P20, P37, and P42 are selected for all considered budget levels. P2 and P10 are valuable with regard to the Forecast Accuracy criterion, and P10 additionally enables project P6. P4 and P42 provide high financial value relative to their cost. P19 and P20 are valuable with regard to the Speed criterion relative to their cost, and P20 is additionally required for P23 as well as synergy S2. Furthermore, P37 is required for execution of P16 and P34. Invariant to the available budget, projects P1, P14, P15, P17, and P35 are never selected. Their overall value is lower than their cost and they do not contribute to the portfolio indirectly through logical interdependencies or synergies. As no aspiration levels warrant the selection of these projects for strategic purposes, these non-profitable projects are not considered in portfolio recommendations.

The decision support system allows decision makers to manually select or exclude specific projects. When a project included in the optimal portfolio for a given budget level is manually excluded, the highest obtainable portfolio contribution (1) taking into account the exclusion will be lower or equal than the contribution when not forcing this project out of the portfolio. The difference in contribution gives the opportunity cost of the project [8]. Likewise, opportunity cost or infeasibility of the integer programming problem (1)–(6) arises when manually selecting a project not included in the optimal portfolio. In order to shed light on the effect of manual adjustments on the portfolio contribution, Fig. 4 presents the percentage loss in portfolio contribution when requiring the selection ($x_j = 1$) or exclusion ($x_j = 0$) of each project $j = 1, \ldots, m$ for budget levels $B = 5, 10, \ldots, 125$. For each project, opportunity cost due to manual selection are indicated by an upward-facing arrow, while opportunity cost due to manual exclusion are represented by a downward-facing arrow. If the manual selection of a project causes the integer programming problem to become infeasible, a cross is given in the chart. Across all budget levels, opportunity cost are particularly high when excluding project P2, valuable with regard to the Forecast Accuracy criterion. Due to high project cost, selecting projects P5, P6, and P12 cause infeasibility for low budget levels and opportunity cost for higher budget levels. While for low budget levels $B \leq 10$ opportunity cost are incurred

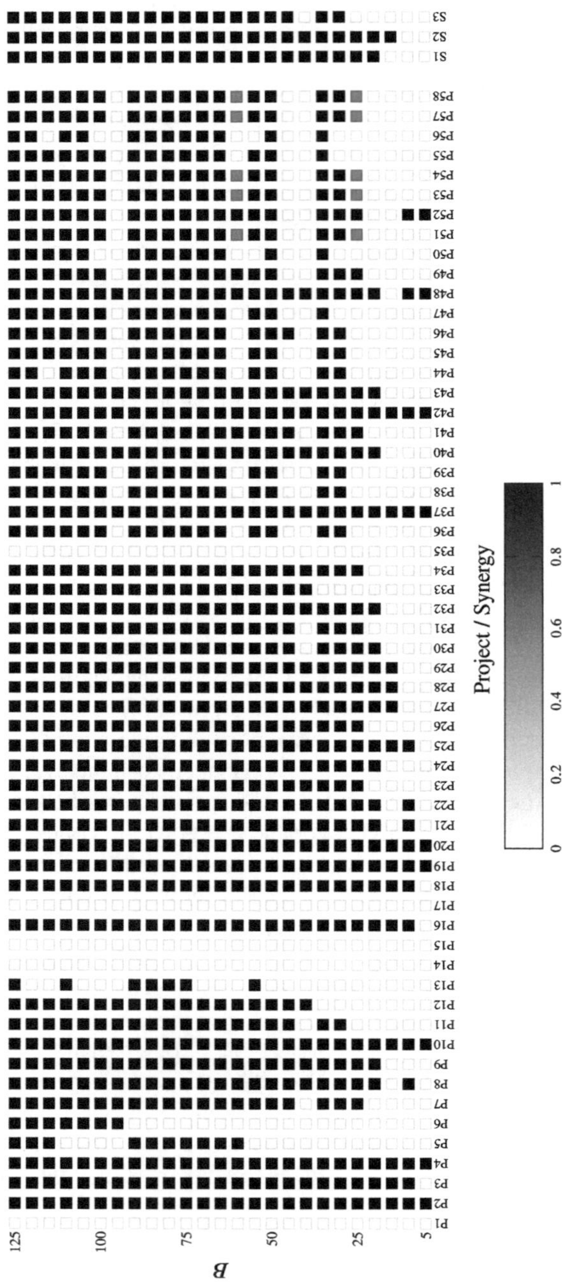

Fig. 3 Relative frequency of project and synergy selection for given budget levels

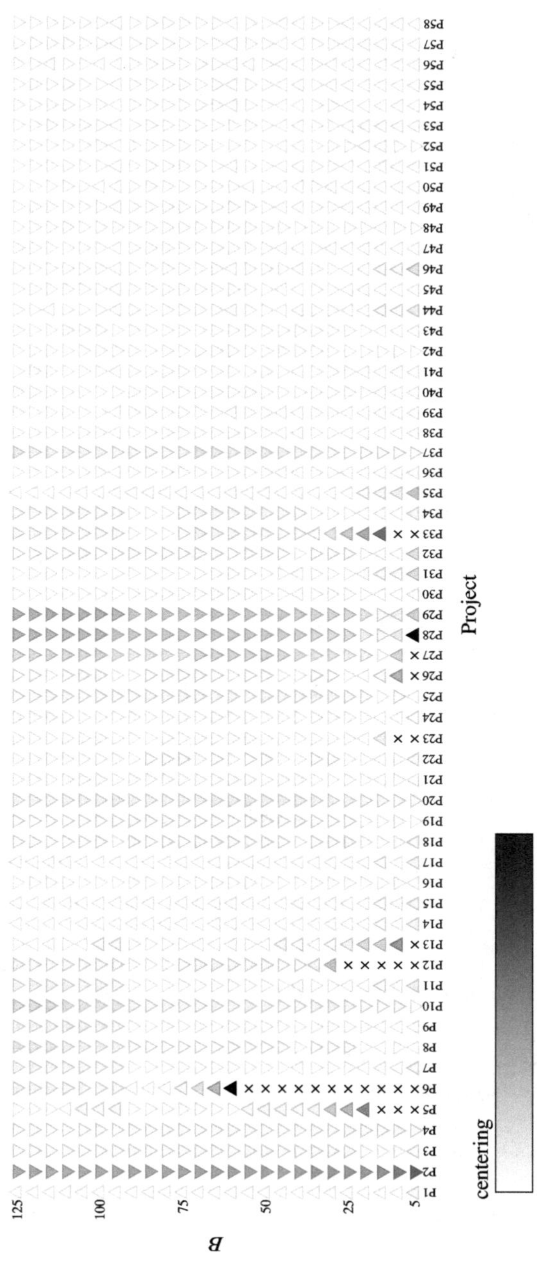

Fig. 4 Percentage loss in portfolio contribution when requiring the selection or exclusion of a project for given budget levels

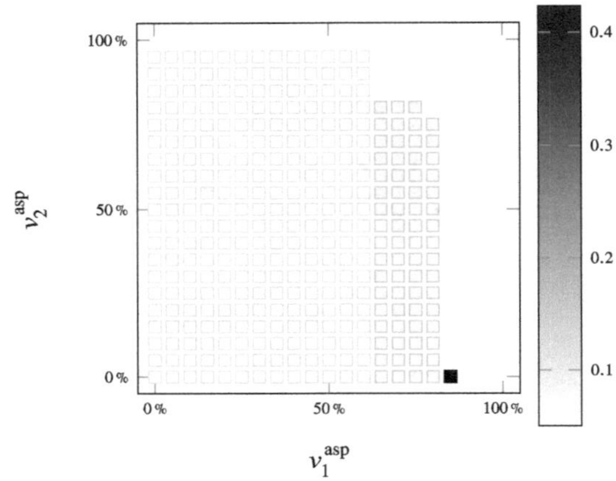

Fig. 5 Percentage loss in portfolio contribution for given aspiration levels v_1^{asp} and v_2^{asp} as percentages of the highest achievable criterion score

when projects P27, P28, and P29 are selected, their exclusion is penalized for higher budget levels. This is due to the fact that projects P28 and P29 are required for the selection of project 27, which in turn shares synergies with projects P9 and P20. Selecting projects P27, P28, or P29 for low budget levels is not cost-efficient as these projects rank among the 15% most expensive projects. On the other hand, project interdependencies prove valuable when sufficient budget is available to select all interdependent projects and utilize synergies.

Finally, we study the effect of aspiration levels on portfolio performance in terms of portfolio contribution (1). Figure 5 reports the loss in portfolio contribution when aspiration levels are imposed compared to the case without aspiration levels. For a fixed budget level $B = 20$ and a fixed aspiration level of financial value $v_3^{asp} = 220$, which equals 95% of v_3^{max}, we vary the aspiration levels of strategic score criteria v_1^{asp} and v_2^{asp}. For a wide range of aspiration levels, decision makers only experience negligible losses in portfolio contribution. Setting aspirations too high causes the problem setting to quickly become infeasible. A trade-off between Speed and Forecast Accuracy arises only at the edge of the space of feasible aspiration levels. Significant losses occur for extreme decision maker preferences, e.g., requiring v_1^{asp} to be greater than 85% of v_1^{max} and setting $v_2^{asp} = 0$. The limited trade-off between Speed and Forecast Accuracy is explained by optimal portfolios accommodating as much strategic value as possible. Projects providing strategic value are preferable as they have a significantly higher (monetized) value-to-cost ratio, 26 for Speed and 46 for Forecast Accuracy on average, than projects providing financial value, which have a ratio of 8, see, e.g., [22] for a discussion of the value-to-cost ratio as an evaluation metric. As portfolios already accommodate high levels of strategic value, aspiration levels are easily achieved without sacrificing much portfolio contribution. These results also hold for higher budget levels and different aspiration levels for financial value.

6 Conclusions and Future Research

This chapter has provided a description of a decision support system for selecting supply chain improvement projects. We have detailed the embedded integer programming model, which is considering project interdependencies and multiple objectives. Thereby, project interdependencies are modeled such that real-world problem instances can be solved in short time to allow the interaction of the user with the decision support system.

Solving a case study with real-world data and undertaking sensitivity analyses has generated a number of interesting insights. First, for varying budget levels subset of projects can be identified which are always or never in the portfolio. This is helpful in the decision analysis as these projects can be included and excluded, respectively, allowing to concentrate the further analysis on the remaining projects. Second, forcing non-selected projects into the portfolio or selected projects out of the portfolio allows valuing projects via opportunity costs. Third, by the variation of aspiration levels, the impact of the considered objectives on the portfolio value can be assessed.

Another important takeaway is the perception of the decision support system by the decision makers. After the first supported session, an initial survey was conducted with three key stakeholders of the decision support system. The decision makers expressed great satisfaction and were eager to employ it in future decision making sessions. Furthermore, the decision makers regarded the decision support system as an important tool for the decision making process and expressed confidence in the system improving the quality of the project portfolio. However, they emphasized a need for proper system documentation and user training in order to get accustomed to the system interface as well as the integer programming model embedded in the system. In terms of future improvements to the decision support system, decision makers proposed including additional information regarding both new project proposals as well as previously selected projects. Decision makers also expressed the need to track ongoing projects and potentially adjust the previously established portfolio by terminating ongoing projects. Finally, decision makers voiced discrepancies between the formal project selection process supported by the decision support system and actual project selection behavior as studied in [33]. While projects proposals are evaluated by standardized, quantitative criteria, which build the foundation of the decision recommendations of the decision support system, decision makers also take into account tacit knowledge and managerial experience not covered by the formal process. This aspect of project portfolio selection has been discussed by Loch [21], based on empirical findings obtained at a European technology manufacturer. Loch [21] concludes that "it is too simplistic to hope for an application of general 'best practice' […] processes". He suggests that organizations should develop a mixture of formal and informal processes, which together meet the strategic needs of the organization needs.

References

1. Archer, N., Ghasemzadeh, F.: Project portfolio selection and management. In: Morris, P.W., Pinto, J.K. (eds.) The Wiley Guide to Managing Projects, pp. 237–255. Wiley, Hoboken (2007). https://doi.org/10.1002/9780470172391.ch11
2. Chien, C.F., Huynh, N.T.: An integrated approach for IC design R&D portfolio decision and project scheduling and a case study. IEEE Trans. Semicond. Manuf. **31**(1), 76–86 (2018). https://doi.org/10.1109/tsm.2018.2792783
3. Cooper, R., Edgett, S., Kleinschmidt, E.: Portfolio management for new product development: results of an industry practices study. R&D Manag. **31**(4), 361–380 (2001). https://doi.org/10.1111/1467-9310.00225
4. De Reyck, B., Grushka-Cockayne, Y., Lockett, M., Calderini, S.R., Moura, M., Sloper, A.: The impact of project portfolio management on information technology projects. Int. J. Project Manage. **23**(7), 524–537 (2005). https://doi.org/10.1016/j.ijproman.2005.02.003
5. Dickinson, M., Thornton, A., Graves, S.: Technology portfolio management: optimizing interdependent projects over multiple time periods. IEEE Trans. Eng. Manage. **48**(4), 518–527 (2001). https://doi.org/10.1109/17.969428
6. Düzgün, R., Thiele, A.: Robust optimization with multiple ranges: theory and application to R&D project selection (2010). https://doi.org/10.1287/ics.2015.0008
7. Fliedner, T., Liesiö, J.: Adjustable robustness for multi-attribute project portfolio selection. Eur. J. Oper. Res. **252**(3), 931–946 (2016). https://doi.org/10.1016/j.ejor.2016.01.058
8. Ghasemzadeh, F., Archer, N., Iyogun, P.: A zero-one model for project portfolio selection and scheduling. J. Oper. Res. Soc. **50**(7), 745–755 (1999). https://doi.org/10.1057/palgrave.jors.2600767
9. Golabi, K., Kirkwood, C.W., Sicherman, A.: Selecting a portfolio of solar energy projects using multiattribute preference theory. Manage. Sci. **27**(2), 174–189 (1981). https://doi.org/10.1287/mnsc.27.2.174
10. Grushka-Cockayne, Y., Reyck, B.D., Degraeve, Z.: An integrated decision-making approach for improving European air traffic management. Manage. Sci. **54**(8), 1395–1409 (2008). https://doi.org/10.1287/mnsc.1080.0878
11. Gurgur, C.Z., Morley, C.T.: Lockheed martin space systems company optimizes infrastructure project-portfolio selection. Interfaces **38**(4), 251–262 (2008). https://doi.org/10.1287/inte.1080.0378
12. Hall, D., Nauda, A.: An interactive approach for selecting IR&D projects. IEEE Trans. Eng. Manage. **37**(2), 126–133 (1990). https://doi.org/10.1109/17.53715
13. Heidenberger, K., Stummer, C.: Research and development project selection and resource allocation: a review of quantitative modelling approaches. Int. J. Manag. Rev. **1**(2), 197–224 (1999). https://doi.org/10.1111/1468-2370.00012
14. Hu, Q.J., Szmerekovsky, J.: Project portfolio selection: a newsvendor approach. Decis. Sci. **48**(1), 176–199 (2016). https://doi.org/10.1111/deci.12214
15. Keisler, J.: Value of information in portfolio decision analysis. Decis. Anal. **1**(3), 177–189 (2004). https://doi.org/10.1287/deca.1040.0023
16. Killen, C.P., Kjaer, C.: Understanding project interdependencies: the role of visual representation, culture and process. Int. J. Project Manage. **30**(5), 554–566 (2012). https://doi.org/10.1016/j.ijproman.2012.01.018
17. Kleinmuntz, D.N.: Resource allocation decisions. In: Edwards, W., Miles, R.F., von Winterfeldt, D. (eds.) Advances in Decision Analysis, pp. 400–418. Cambridge University Press, Cambridge (2007). https://doi.org/10.1017/cbo9780511611308.021
18. Kolisch, R., Brunner, J.O., Fliedner, T.: Project portfolio planning with project interdependencies and multiple criteria: a practical application. In: Demeulemeester, E.L., Herroelen, W. (eds.) PMS 2012 - the 13th International Conference on Project Management and Scheduling (2012)

19. Liesiö, J., Mild, P., Salo, A.: Preference programming for robust portfolio modeling and project selection. Eur. J. Oper. Res. **181**(3), 1488–1505 (2007). https://doi.org/10.1016/j.ejor.2005.12.041
20. Liesiö, J., Salo, A.: Scenario-based portfolio selection of investment projects with incomplete probability and utility information. Eur. J. Oper. Res. **217**(1), 162–172 (2012). https://doi.org/10.1016/j.ejor.2011.08.025
21. Loch, C.: Tailoring product development to strategy: case of a European technology manufacturer. Eur. Manag. J. **18**(3), 246–258 (2000). https://doi.org/10.1016/s0263-2373(00)00007-4
22. Lourenço, J.C., Morton, A., Bana e Costa, C.A.: PROBE—a multicriteria decision support system for portfolio robustness evaluation. Decis. Support. Syst. **54**(1), 534–550 (2012). https://doi.org/10.1016/j.dss.2012.08.001
23. Martinsuo, M., Korhonen, T., Laine, T.: Identifying, framing and managing uncertainties in project portfolios. Int. J. Project Manage. **32**(5), 732–746 (2014). https://doi.org/10.1016/j.ijproman.2014.01.014
24. Marx, F., Mayer, J.H., Winter, R.: Six principles for redesigning executive information systems—findings of a survey and evaluation of a prototype. ACM Trans. Manag. Inf. Syst. **2**(4), 1–19 (2011). https://doi.org/10.1145/2070710.2070717
25. Mohagheghi, V., Mousavi, S.M., Antuchevičienė, J., Mojtahedi, M.: Project portfolio selection problems: a review of models, uncertainty approaches, solution techniques, and case studies. Technol. Econ. Dev. Econ. **25**(6), 1380–1412 (2019). https://doi.org/10.3846/tede.2019.11410
26. Morton, A., Keisler, J.M., Salo, A.: Multicriteria portfolio decision analysis for project selection. In: Greco S., Ehrgott M., Figueira J. (eds.) Multiple Criteria Decision Analysis, pp. 1269–1298. Springer, New York (2016). https://doi.org/10.1007/978-1-4939-3094-4_28
27. Panadero, J., Doering, J., Kizys, R., Juan, A.A., Fito, A.: A variable neighborhood search simheuristic for project portfolio selection under uncertainty. J. Heuristics **26**(3), 353–375 (2018). https://doi.org/10.1007/s10732-018-9367-z
28. Phillips, L.D., e Costa, C.A.B.: Transparent prioritisation, budgeting and resource allocation with multi-criteria decision analysis and decision conferencing. Ann. Oper. Res. **154**(1), 51–68 (2007). https://doi.org/10.1007/s10479-007-0183-3
29. Roland, J., Figueira, J.R., Smet, Y.D.: Finding compromise solutions in project portfolio selection with multiple experts by inverse optimization. Comput. Oper. Res. **66**, 12–19 (2016). https://doi.org/10.1016/j.cor.2015.07.006
30. Sampath, S., Gel, E.S., Fowler, J.W., Kempf, K.G.: A decision-making framework for project portfolio planning at Intel corporation. Interfaces **45**(5), 391–408 (2015). https://doi.org/10.1287/inte.2015.0809
31. Sampath, S., Gel, E.S., Kempf, K.G., Fowler, J.W.: A generalized decision support framework for large-scale project portfolio decisions. Decis. Sci. (2021). https://doi.org/10.1111/deci.12507
32. Santhanam, R., Kyparisis, G.J.: A decision model for interdependent information system project selection. Eur. J. Oper. Res. **89**(2), 380–399 (1996). https://doi.org/10.1016/0377-2217(94)00257-6
33. Schiffels, S., Fliedner, T., Kolisch, R.: Human behavior in project portfolio selection: insights from an experimental study. Decis. Sci. **49**(6), 1061–1087 (2018). https://doi.org/10.1111/deci.12310
34. Stewart, T.J.: Multiple objective project portfolio selection based on reference points. J. Bus. Econ. **86**(1–2), 23–33 (2016). https://doi.org/10.1007/s11573-015-0797-x
35. Stummer, C., Heidenberger, K.: Interactive R&D portfolio analysis with project interdependencies and time profiles of multiple objectives. IEEE Trans. Eng. Manage. **50**(2), 175–183 (2003). https://doi.org/10.1109/tem.2003.810819

36. Tavana, M., Sodenkamp, M.A.: A fuzzy multi-criteria decision analysis model for advanced technology assessment at Kennedy space center. J. Oper. Res. Soc. **61**(10), 1459–1470 (2010). https://doi.org/10.1057/jors.2009.107
37. Ward, S., Chapman, C.: Transforming project risk management into project uncertainty management. Int. J. Project Manage. **21**(2), 97–105 (2003). https://doi.org/10.1016/s0263-7863(01)00080-1

Index

A

Ant colony optimization, 58
  non-outranked, 26
Artificial bee colony, 137, 139, 141, 146

B

Bayesian network, 27, 160
Branch and bound, 4, 161, 172, 179, 183

C

Capability option, 89, 90, 96, 105, 121
Conventional optimization, 4, 22, 37–39

D

Decision support system, 193, 195, 196, 198, 199
Differential evolution, 29, 37, 38, 56, 95, 107
  multi-objective, 137, 141
Dynamic programming, 133, 161

E

ELECTRE, 28, 31, 80, 82
Estimation of distribution algorithm, 27
Evolutionary algorithm, 10, 26–28, 37, 78, 81, 84, 146, 196
  multi-objective, 26, 28, 78
  multi-objective, based on decomposition, 26, 137, 141
  strength Pareto, 26, 28
  strength Pareto (2), 28, 137, 141
Evolutionary computation, 5, 22, 29, 127
Evolutionary programming, 56
Evolution strategy, 52
  covariance matrix adaptation, 27, 53

F

Firefly algorithm, 145
Future force design, 13, 89, 90, 121
Fuzzy logic, 137, 141
Fuzzy programming, 137, 139, 141

G

Genetic algorithm, 29, 37, 41, 43, 94, 106, 133, 136, 137, 139, 141, 195, 196
  biased random-key, 29, 95, 107
  binary, 45
  hybrid multi-criteria sorting, 26
  micro, 59
  multi-objective, 147
  non-dominated sorting (II), 26, 28, 29, 82, 137, 139, 141
  non-dominated sorting (III), 26
  non-outranked sorting, 26, 82
    interval, 82
  random-key, 95
Genetic programming, 38, 137, 139, 141
Goal programming, 22, 25, 29, 32, 150

H

Heuristic, 32, 59, 68, 69, 89, 91, 102, 108, 121, 138, 161, 181, 196

I

Interdependencies, 11, 19, 20, 90, 164, 193, 194, 197, 201, 205, 208, 209

K. R. Harrison et al., *Evolutionary and Memetic Computing for Project Portfolio Selection and Scheduling*, Adaptation, Learning, and Optimization 26,
https://doi.org/10.1007/978-3-030-88315-7

Invasive weed optimization, 137, 141

K
K-means clustering, 27, 31
Knapsack problem, 13, 16, 32, 68, 92, 160, 168, 183
  inverse, 187
  multidimensional, 92, 93, 101
  multiple, 161, 168, 186
  multiple-choice, 93, 100, 101
  multiple-choice multidimensional, 93, 101, 102
  quadratic, 187
  set-union, 93, 101, 102

L
Learning classifier systems, 38

M
Markov chain, 138
Memetic algorithm, 10, 32, 37, 38, 59, 161
Monte Carlo simulation, 28, 146, 196
Multi-attribute value theory, 27
Multi-criteria decision making, 28, 78, 80, 166, 194
Multi-objective optimization, 2, 26, 29, 39, 40, 133, 138, 148
Multiple objectives, 18, 22, 29, 39, 66, 161, 193–195, 209

O
Opportunity cost, 205, 209

P
Particle swarm optimization, 58, 136, 137, 141, 145
  multi-objective, 137, 141
Preference disaggregation analysis, 26
Programming
  integer, 4, 41, 194–199, 201
  linear, 4, 27, 28, 31, 41, 126, 136, 144, 146, 160, 161, 172, 174, 179, 180, 186
  mixed-integer, 27, 31, 136, 172, 179
  non-linear, 4, 136, 144, 195
  quadratic, 126, 144
PROMETHEE, 67, 80

R
Reinforcement learning, 28
Robust optimization, 126

S
Simulated annealing, 58
Stochastic optimization, 133

T
Tabu search, 59, 137, 141

W
Whale optimization algorithm, 145